Testlet Response Theory and Its Applications

The measurement models employed to score tests have been evolving over the past century from those that focus on the entire test (true score theory) to models that focus on individual test items (item response theory) to models that use small groups of items (testlets) as the fungible unit from which tests are constructed and scored (testlet response theory, or TRT).

In this book, the inventors of TRT trace the history of this evolution and explain the character of modern TRT. Written for researchers and professionals in statistics, psychometrics, and educational psychology, the first part of the book offers an accessible introduction to TRT and its applications. The rest of the book is a comprehensive, self-contained discussion of the model couched within a fully Bayesian framework. Its parameters are estimated using Markov chain Monte Carlo procedures, and the resulting posterior distributions of the parameters yield insights into score stability that were previously unsuspected. The authors received the National Council on Measurement in Education award for scientific contribution to a field of educational measurement for this work.

HOWARD WAINER is a Distinguished Research Scientist for the National Board of Medical Examiners and Adjunct Professor of Statistics at the Wharton School of the University of Pennsylvania. Among his many honors, he received the Educational Testing Services' Senior Scientist award and is a Fellow of the American Statistical Association. This is his fifteenth book.

ERIC T. BRADLOW is the K. P. Chao Professor; Professor of Marketing, Statistics, and Education; and Academic Director of the Wharton Small Business Development Center at the Wharton School of the University of Pennsylvania. Before joining the Wharton faculty, he worked in the Corporate Marketing and Business Research Division at the DuPont Corporation and in the Statistics and Psychometrics Research Group at the Educational Testing Service. Bradlow was recently named a Fellow of the American Statistical Association.

XIAOHUI WANG is an Assistant Professor in the Department of Statistics at the University of Virginia. She worked as a Principal Data Analyst for three years in the Division of Data Analysis and Research Technology at the Educational Testing Service. She has twice received the National Council on Measurement in Education Award for Scientific Contribution to a Field of Educational Measurement.

Testlet Response Theory and Its Applications

HOWARD WAINER
National Board of Medical Examiners

ERIC T. BRADLOW
University of Pennsylvania

XIAOHUI WANG
University of Virginia

To Sandip Sinharay
With my gratitude for
your help & my best wishes —
Howard Wainer
Princeton
Nov 15, 2016

CAMBRIDGE UNIVERSITY PRESS
Cambridge, New York, Melbourne, Madrid, Cape Town, Singapore, São Paulo

Cambridge University Press
32 Avenue of the Americas, New York, NY 10013-2473, USA

www.cambridge.org
Information on this title: www.cambridge.org/9780521862721

First published 2007

Printed in the United States of America

A catalog record for this publication is available from the British Library.

Library of Congress Cataloging in Publication Data

Wainer, Howard.
Testlet response theory and its applications / Howard Wainer, Eric T.
Bradlow, Xiaohui Wang.
p. cm.
Includes bibliographical references and index.
ISBN-13: 978-0-521-86272-1 (hardback)
ISBN-10: 0-521-86272-8 (hardback)
ISBN-13: 978-0-521-68126-1 (pbk.)
ISBN-10: 0-521-68126-X (pbk.)
1. Examinations – Scoring. 2. Bayesian statistical decision theory.
I. Bradlow, Eric T. II. Wang, Xiaohui III. Title.
LB3060.77.W35 2007
371.26 – dc22 2006030231

ISBN 978-0-521-86272-1 hardback
ISBN 978-0-521-68126-1 paperback

To Linda, Laura, and Weiqiang

Contents

Preface

This book describes the outcome of a research program that began in the mid-1980s. In the more than two decades since, many debts, both intellectual and financial, have been incurred. We are delighted to have the opportunity to acknowledge the help and express our gratitude.

First, organizational thanks are due to the National Board of Medical Examiners and its president, Donald Melnick, who has supported this work for the past five years. Senior Vice President Ronald Nungester and Associate Vice President Brian Clauser's enthusiasm for this work is especially appreciated.

Second, our gratitude to the Educational Testing Service (ETS), which employed all of us during the critical period in which the Bayesian version of testlet response theory was being birthed, is equally sincere. Henry Braun, then Vice President of Research at ETS, was a crucial voice in committees that provided support for this project. His enthusiasm and wise counsel were appreciated then and now. The funding organizations that provided support were the Graduate Record Board, the TOEFL Board, the College Board, the Joint Scientific Research and Development Committee, the Law School Admissions Council, and the research budget of ETS. We are also grateful to Kurt Landgraf, president of ETS, whose wise leadership made continued support of basic research financially possible.

Intellectual debts are harder to keep track of than financial ones, and if we have omitted anyone we hope that the lapse will be recognized as one of memory and not ingratitude. Our thanks, in more-or-less chronological order, go to

Gerald Kiely, who spent a summer at ETS as an intern and helped with the birth of the testlet concept.

Paul Rosenbaum, who before he went on to fame as the world's leading expert on observational studies occupied many hours discussing the use of item bundles as a viable practical solution to various pressing problems in modern testing.

His proof of the plausible conditional independence of bundles provided important supporting evidence for the concept.

David Thissen, whose collaboration on the use of a polytomous IRT model to score testlet-based tests was key to making the ideas easily operational and who helped us show how its use affects reliability.

Charles Lewis, for his penetrating analyses of test construction and scoring as well as his gentle insistence on continued exploration to make whatever we had better.

Stephen Sireci, who as a summer intern was the operational force behind the development of the initial testlet methods for studying DIF and for proving the importance of the testlet concept by showing the extent to which ignoring local dependence upwardly biases estimates of test reliability.

Robert Mislevy, who pointed out the critical importance of Robert Gibbons' bi-factor test scoring model as a more general testlet model, and who has made himself available countless times for discussions of issues both technical and logical; most recently on some vexing questions on the identifiability of parameters when covariates are used.

Zuru Du, another ETS summer intern, who came to Princeton in search of a thesis topic and left with the task of expanding the 2-PL testlet model to accommodate guessing. His success at doing this earned him his PhD and subsequent awards, as well as our thanks.

Cees Glas, who demonstrated that maximum likelihood estimation is not yet obsolete by using it to fit the 3-PL testlet model. He also wrote the code that did it by lunchtime.

The enjoyment, as well as the value, of writing a book loses more than just its patina of scholarly accomplishment if there is no prospect of it ever being published. Thus, our gratitude to Cambridge University Press and mathematics editor Lauren Cowles for ready acceptance and encouragement is much more than pro forma. When we first discussed the pros and cons of Cambridge doing the book Lauren pointed out that they were the outlet Newton chose to publish his *Principia.* That was enough of a recommendation for us.

Any project whose execution spanned decades accumulates debts to many others whose help was important. Most important are Elizabeth Brophy, Editha Chase, and Martha Thompson, whose work ethic and keen sense of organization kept everything in order.

Our gratitude to Navdeep Singh and the rest of his production staff for their help in making this book the best it could be.

The writing of this book was guided, in part, by the wisdom of two great physicists – Albert Einstein, who pointed out that "everything should be as simple as possible; but no simpler," and J. Robert Oppenheimer, who in his class on quantum mechanics told his students that "I may be able to make it clearer, but I can't make it simpler."

The book begins conceptually, requiring no more mathematics than simple algebra and some basic statistical concepts, but it ends up much more technical. Part I (Chapters 1–6) is no more complex than any modern book on test theory, although there is an occasional integral.

Part II (Chapters 7–12), where we introduce and develop Bayesian testlet response theory, requires more. Here the technology of Bayesian methods with its requirement of distribution theory and an understanding of how a Markov process can eventually converge to a stationary result is an area that is relatively new and may require slower going. We have included a tutorial in these methods at the very end of the book (Chapter 15) that can be helpful in two ways. It provides an introduction that may be helpful by itself – its glossary can aid those for whom some of the language is unfamiliar – and its reference list can point interested readers to further details.

Part III (Chapters 13–15) contains two applications of the new model and its associated technology. These can be read without a full understanding of the mathematics and hopefully will whet the reader's appetite enough to warrant gaining the technical expertise for a deeper understanding.

Last, we have prepared a computer program (SCORIGHT) that can do everything we describe here. It is available free from us. You can download SCORIGHT's users manual from www.cambridge.org/9780521681261 as well as a permission-to-use form. To get a copy of the current version of SCORIGHT merely print out the permission form, sign it, and mail it (with your email address) to:

Howard Wainer
Distinguished Research Scientist
National Board of Medical Examiners
3750 Market Street
Philadelphia, Pennsylvania 19104

PART I

Introduction to Testlets

1

Introduction

To count is modern practice, the ancient method was to guess.
Samuel Johnson

The subject of this book is testing. The origin of testing lies within a magical epiphany. Almost four thousand years ago, an anonymous functionary was focused on a task set before him by the emperor of China. He needed a procedure to choose wisely from among all applicants for government positions. Perhaps, utilizing a model made famous by Newton three millennia later, he was lying under a tree when the fundamental principle of testing struck him:

> *A small sample of behavior, gathered under controlled circumstances, can be predictive of performance in a much broader set of uncontrolled circumstances.*

Testing is an ingenious solution to a practical problem. In an ideal world, before choosing among candidates for some position, we would like to know how they would perform in that position. Obviously, if we could place each of them in the position for a long enough period of time we would have direct evidence, but such a solution is usually not a practical possibility. But, even in a situation in which we had, say a year of trial, we would still not know for sure how they would perform after two years, or three, or more. In this instance, however, we would be pretty convinced of the validity of the task utilized to make the decision – it is, after all, the same job. But if the tasks on the job vary, the one-year trial period may not be long enough. In addition, one year may not be long enough to allow us to be sure – our judgments about the person's performance on each of the components of the position may not be as reliable as we would like. No matter what we do, we must make some sort of extrapolating inference.

The formal Chinese Civil Service tests used during the Chan Dynasty (1115 B.C.) were but a continuation of practices that were already a thousand years old. These were job sample tests assessing proficiency in archery, arithmetic, horsemanship, music, writing, and various social skills. The candidate's total test score was a summary of his scores on these subtests, or what has come to be called testlets. The Chinese testing enterprise, as innovative and clever as it was, did not have a formal theory for how to combine the candidate's performance on these several testlets. The development of such a theory lay almost three thousand years in the future. This book describes a theory of testlets and the pathway that was traversed in its development.

We have found that an often-useful strategy in any problem of experimental design is to start our thoughts with an optimal design, without regard to practical constraints, and then shrink back to a solution within the ever-present practical limits. By proceeding in this manner, exactly what must be assumed to be true without evidence is made explicit. It doesn't take too much of a stretch of one's imagination to conceive of that ancient Chinese functionary, whose very bones are now long dust, reasoning in this manner. He must have wondered:

i. What were the key tasks that the selected persons would have to perform?
ii. How could those tasks be represented?
iii. How can each examinee's performance be rated?
iv. How accurate is this artificial set of tasks at predicting future performance? And, if he was as clever as we reckon he must have been,
v. How accurate would this artificial set of tasks be at predicting future performance if the task constructor made modest errors in each of these steps?

These ancient issues remain the key questions that face modern testing. The first step in preparing any test is deciding what information about the examinees the users of the test scores will want to know. If we are trying to choose among applicants to college, we might want to know something about the applicants' scholarly ability, their maturity, and their grit; in short, their ability to do the myriad of tasks it takes to succeed in a collegiate environment. We probably would not be much interested in their height, weight, or eye color.

The second question is how are the tasks that we assign related to the variables of interest (construct validity). Thus, the question of primary interest might be "how well can this person write descriptive prose?" After we agree to this, we then examine the various kinds of tasks that we could ask that would provide evidence about the person's writing ability. We could ask the person to write some descriptions; we might ask the person to answer some verbal analogy items; we might even ask the person to run 100 meters as fast as she can. Or we could construct a test that was a combination of some or all of these. Obviously, each of these tasks bears a differential connection to the original question. In practice, the choice of representation depends on many variables, some scientific and some practical.

The third question can be restated into "how can this test be scored?" At this point, it is important to make a distinction between an item and a question. An item has a surface similarity to a question, but "item" is a term with a very specific technical meaning. The term "question" encompasses any sort of interrogative:

Where's the bathroom?
What time is it?
Do you come here often?
What's your sign?

These are all questions, but they should not be confused with items. For a question to be called an item, it needs an additional characteristic:

There must be a scoring rule

Tests can be scored in many ways. Traditionally, "points" were allocated according to some sort of rule. When it was possible, items were scored right or wrong and the overall score was the number or percentage correct. With items like essays that could not be divided neatly into binary categories, raters would follow some sort of scoring rubric to allocate points. In the latter half of the 20th century, formal statistical models were introduced to take some of the guesswork out of this procedure (see Thissen & Wainer, 2001, for a detailed description of test scoring).

The fourth question, accuracy of prediction, is much more complex than might first appear. It is the fundamental issue of test validity and is composed of many pieces. Taking the simplistic view of validity as just a question of predictive accuracy assumes that one can determine a single criterion. This is rarely, if ever, the case. Moreover, even when it is the case, there is rarely a single way to characterize that criterion. Few would deny that medical schools should admit those applicants who are likely to become the best physicians. But how do you measure that criterion?

In addition, most tests have multiple purposes, and the validity of the test's score varies with the purpose. Entrance tests for college are often aggregated and then used as measures of the quality of precollegiate education. While it is reasonable to believe that the same test that can compare individuals' college readiness also contains information about their high school education, a special purpose test for the latter (probably focusing more on specific subject matter – as with the Scholastic Assessment Test (SAT) II) seems like a better option. But preparing separate specific tests for each of many closely allied purposes is often impractical, and hence we usually approximate the test we want with the test we have.

And, finally, the fifth question, the question of robustness, which is the driving issue behind the developments described in this book: How well is any particular test score going to work if the various assumptions that must be made in its construction are not quite correct? Obviously, this is a quantitative question. If the inferences being made are inexact, the testing instrument can afford small infidelities. Millennia past, in Han Dynasty China, the selection process that utilized their tests was so extreme that anyone who survived the screening had to be very able indeed. There were thousands of candidates for every opening, and so even though there were many worthy candidates who were turned down, the ones who were chosen were surely exceptional. The screening test did not have to be great to find extraordinary candidates.

This same situation exists today. The screening for National Merit Scholars begins with about 1.5 million candidates and eventually winnows down to 1,500 winners. The tests that accomplish this do not have to be very valid – just a little – in order for the winners to be special. Of course there are many, many worthy candidates who are not chosen, but no one could argue that any of the winners were not worthy as well. Herman Wold, a member of the committee that chooses Nobel Prize winners in economics, once explained that choosing winners was easy, because in the tails of the distribution data points are sparse and there are big spaces between them. When that is the situation, the choice algorithm, and the evidence that it uses, does not have to be precise.

Testing has prospered over the nearly four millennia of its existence because it offered a distinct improvement over the method that had preceded it.

But as time has passed and test usage increased, the demands that we have made on test scores have increased, as has the "fineness" of the distinctions that we wish to make. As this has happened, tests and how they are scored have improved. These improvements have occurred for three principal reasons:

1. The demands made on the test scores have become more strenuous.
2. We have gathered more and more evidence about how well various alternatives perform.
3. Our eyes have become accustomed to the dim light in those often dark and Byzantine alleys where tests and psychometrics live.

Let us examine each of these in more detail:

Tests have three principal purposes: (i) measurement, (ii) contest, and (iii) prod. For the first 3,950 years of their existence, tests were principally contests. Those with high enough scores won (got the job, received the scholarship, were admitted) and everyone else lost (didn't get the job, didn't get the scholarship, were rejected). For some of that time, tests also served as prods: "Why are you studying math tonight?" "I have a test tomorrow." In the modern world, with the enormous prevalence of tests, "prod" is still often the primary purpose. A test as a measurement tool is the most modern usage. It was only in the past century that a statistical theory of test scores was developed to provide a rigorous methodology of scoring that allows us to think of test scores as measurement in a way that was more than rudimentary in character (Gulliksen, 1950/1987 Lord & Novick, 1968; Thissen & Wainer, 2001). These formal developments occurred in response to users of tests who wanted to know more than merely who won. Test scores became one measure of the efficacy of the educational system. They were asked to express the effectiveness of children's learning progress. And, by comparing scores of different subgroups within society, they were used to express the fairness of that society.

As the usage of tests for these many purposes increased, evidence accumulated that supported some previously held notions, contradicted some others, and revealed some things that were never suspected (e.g., the precise character of the heritability of some mental traits (Bock & Kolakowski, 1973)). The multiple-choice item was developed for ease of scoring, but as evidence accumulated, it was discovered that it could provide valid evidence of skills and knowledge well beyond what its inventors could originally have even dreamed. One example of this (there are many others) is revealed in the study of writing ability (Wainer, Lukele, & Thissen, 1994). Few would think that multiple-choice items were a sensible way to measure writing ability, yet in fact when three 30-minute tests were given, in which two were essays and one was a mixture of verbal analogies and other typical multiple-choice verbal items, it was found that examinees' scores on the multiple-choice test correlated more highly with either of the two essay test scores than the essay test scores did with one another! What this means is that if you want to know how someone will do on some future essay exam, you can predict that future essay score better from the multiple-choice exam than you could from an essay exam of the same length.

The reason for this remarkable result illuminates the value of multiple-choice exams. Any test score contains some error. This error has several components. One component is merely random fluctuation; if the examinee took the same test on another day he might not respond in exactly the same way. A second component is associated with the context of the exam; if one is asked to write an essay on the American Civil War, it might not be as good as one on World War II, because the examinee might just know more about the latter topic. But if the test is supposed to measure writing ability, and not subject matter knowledge, this variation is non–construct-related variance, and hence in this context is thought of as error. A third component of error is scoring variance; essays are typically scored by human judges, and even the best-trained judges do not always agree; nor would they provide the same score if they were to rescore the exam (Bradlow & Wainer, 1998). Hence, an essay's score might vary depending on who did the scoring. A fourth component is construct bias; the test might not be measuring the intended thing. So, for example, suppose we are interested in measuring a teenager's height, but all we have is a bathroom scale. We could use the child's weight to predict her height (taller children are generally heavier), but the prediction would probably contain a gender bias because teenage girls typically weigh less than boys of the same height. Hence boys would probably be underpredicted, and girls overpredicted.

A multiple-choice test of writing may contain a bias that is not in an essay test, because it is measuring something different from writing, but the size of two of the other three components of error variation is almost surely smaller. There is no judge-to-judge scoring variation on a multiple-choice exam. And, because a 30-minute multiple-choice exam can contain 30 or so items, a broad range of topics can be spanned, making it unlikely that large idiosyncratic topic effects would exist. Whether there is a difference in day-to-day variation in student performance on any specific essay versus a multiple-choice test is a topic for experiment, but is not likely to be substantial.

So now the answer is in sight. The multiple-choice writing test predicts future essay scores better than does an essay score of the same length because the construct bias is small relative to the other sources of error. Of course, the other errors on essay tests can be reduced by increasing the number of essays that are written; by making each one longer; and by having many judges score each one. But in most circumstances, these options make the test too cumbersome and would make the test too expensive for the examinee if such a policy were implemented. And, even if so, a longer multiple-choice exam might still be better.

Being able to measure the wrong thing accurately as opposed to the right thing badly has been the saving grace of modern testing so long as the wrong thing is not too far wrong. It has also allowed psychometricians to work apparent magic and connect tests on different topics so that they can be placed on comparable scales. This is possible because of the generally high correlation among all categories of ability. Students who are good at math are generally good at science; those who have well-developed verbal skills are also good at history, sociology, and geography. A large number of 20th-century tests have relied heavily on the high covariation among different tests to allow scores on very different tests to have approximately the same meaning. For example, the achievement tests given to prospective college students (SAT II) are in many subjects (e.g., math, chemistry, physics, history, French, Spanish, German, and Hebrew). The scores on each of these tests are scaled from a low of 200 to a high of 800. Users treat the scores as if a 600 on Hebrew means the same as a 600 on French or physics. But how could it be? What does it mean to say that you are as good in French as I am in physics?

One possible way to allow such a use would be to scale each test separately, so that they each had the same mean and standard deviation. But this would be sensible only if we believed that the distribution of proficiency of the students who chose each examination were approximately the same. This is not credible; the average person taking the French exam would starve to death on the Boulevard Raspail, whereas the average person taking the Hebrew exam could be dropped into the middle of Tel Aviv with no ill effects. Thus, if this hypothesis of equal proficiencies were not true, then all we could say is that you are as able relative to the population that took "your" test as I am to the population that took mine. This limits inferences from test scores to only among individuals who took the same form of the test, which is not powerful enough for current demands.

Instead what is done is to lean heavily on the covariation of ability across subjects by (for language tests) using the verbal portion of SAT I as an equating link between the French and Hebrew exams. In a parallel fashion, the mathematical portion of SAT I is used as the common link between the various science exams. The same idea is useful in assessing the prospective value of an advanced placement course for an individual student on the basis of her performance on other tests (Wainer & Lichten, 2000).

This high level of covariation between individual performance on tests, of often substantially different subject matter, led early researchers to suggest that performance was governed by just a single, underlying general ability that came to be called "g." Almost a century of research has shown that although this conjecture was incorrect, it was only barely so (Ree & Earles, 1991, 1992, 1993). The unified structure of proficiency has helped to give tests of widely different characters enough predictive validity to be of practical use.

After the value of mass-administered standardized testing was confirmed with its initial military uses in the first quarter of the 20th century, its use ballooned and now invades all aspects of modern life, from school admissions, to educational diagnosis, to licensure. It has even played an integral role in a U.S. Supreme Court decision of whether or not a convicted murderer was smart enough to be executed (*Atkins v Virginia,* 2002). As testing has become more integral to education and licensing, ways to increase its efficiency have been sought. As we discussed earlier, the movement in the first half of the 20th century toward a multiple-choice format has been one successful improvement. The movement toward computerized test administration at around the beginning of the 21st century has been another.

Computerizing test administration has been done for several purposes. Two reasons that loom especially large are (i) for testing constructs difficult or impossible to test otherwise (e.g., testing proficiency in the use of CAD-CAM equipment) and (ii) for providing a test customized for each examinee. It is the second purpose that is relevant to our discussion at this time. Providing customized exams is the purpose of what has become known as computerized adaptive testing, or CAT (Wainer et al., 2000). In CAT, the test is broken up into smaller units, traditionally items, and testing begins with an item of middling difficulty being presented to an examinee. If the examinee gets the item correct, a more difficult one is presented; if the examinee gets it wrong, an easier one. In this way, the examinee is faced with fewer items that are of little value in estimating his or her ability and the estimation of ability proceeds more quickly to an accurate estimate. The increased efficiency of such a procedure is of obvious value for usages like instructional diagnosis, when an instructor would like to know, with reasonable accuracy, what the student doesn't know. And the instructor would like to know this with as short a test as possible. To build a test with enough items for accurate diagnosis in all areas would make it unwieldy in the extreme. But by making it adaptive, a student could demonstrate mastery quickly and obviate all diagnostic questions meant to ferret out exactly the area of misunderstanding. Adaptive tests can also provide greater precision of estimation within a fixed amount of testing time. Current experience suggests that CAT can yield the same precision as a fixed-length test that has about twice the number of items.

For these spectacular outcomes to occur, CAT requires an enormous bank of items for the testing algorithm to choose from as it builds the test. It also needs a measurement model (a theoretical structure) that can connect the various individual tests in a way that allows us to make comparisons among examinees who may have taken individualized test forms with few, or even no, items in common. The formal test scoring models that were originally developed are grouped under the general label "true score theory" and focused on the entire test as the unit of observation. This approach will be discussed in Chapter 2. True score theory carried testing a long way but was insufficient for situations like CAT in which test forms are constructed on the fly. In this situation, the "test" was too large a unit and instead the individual item was considered the building block of the theory, which was thus named "item response theory," or IRT. We will elaborate on IRT in Chapter 3.

IRT is a family of models that try to describe, in a stochastic way, what happens when an item meets an examinee. To do this, it makes some assumptions about the character of the response process. These assumptions range from models that have very restrictive assumptions to others for which those assumptions are relaxed considerably. But until recently, all IRT models had conditional local independence as one of their primary requirements. What this means is that an examinee's performance on any item depends only on the examinee's ability and that item's characteristics, and that knowledge of the examinee's performance on another item (or items) does not add any further information. Extensive evidence has shown that this assumption was usually accurate enough for the sort of tests being built and the inferences being made. But it did force some limitations on test construction that were unfortunate. For example, local independence did not hold for typical reading comprehension items that consisted of a reading passage and a set of four to six associated items. The U.S. military in the CAT version of their entrance test (CAT ASVAB – Armed Services Vocational Aptitude Battery) solved the problem by using only a single item with each passage. This was far from an optimal solution since it was inefficient to use up a considerable amount of examinee time in reading the passage, and then extract only one bit of information from it. Reducing the length of each passage considerably ameliorated this, but it was then discovered that the construct being measured with this new formulation was more heavily tilted toward vocabulary and away from the comprehension of descriptive prose. The Educational Testing Service on many of its tests employed an alternative strategy. It limited the number of items per passage to four to six and assumed conditional independence, hoping that it would be accurate enough; and for many of the tests it was.

But as evidence accumulated and tests were being asked to enable more delicate inferences, it became clear that the assumption of local independence was often untenable. Test developers were finding that with increasing frequency it was important to build a test out of units that were larger than a single item. Units that might be a statistical graphic or table followed by a number of questions; a passage in a foreign language presented orally and accompanied by a number of questions; a musical piece played and then several questions asked. The constructs that were being measured required such aggregations of items, which by their very nature, were not conditionally independent. It soon became clear that a measurement model was required in which the fungible unit of test construction was smaller than the whole test yet larger than a single item. Wainer and Kiely (1987) argued for such a model, proposing the *testlet* as the unit of test construction. The testlet was meant as an extremely flexible unit.

While it was originally proposed as an aggregation of items that were grouped together to act as a unit, the concept could encompass testlets with but a single item, or, at the other extreme, a test could be made up of a single testlet. Once the concept of the testlet was suggested, a number of alternative psychometric models were proposed.

It is these models and the uses that can be made of them that are the subjects of this book.

Questions

1. What is a test?
2. What are the three uses of tests?
3. How is an item different from a question?
4. Why does large-scale testing rely on the concept of general intelligence?
5. Why has modern testing moved toward computerized administration?
6. What is a testlet?
7. What has been the driving force behind the invention of the testlet?

References

Bock, R. D., & Kolakowski, R. (1973). Further evidence of sex-linked major-gene influence on human spatial visualizing ability. *American Journal of Human Genetics, 25,* 1–14.

Bradlow, E. T., & Wainer, H. (1998). Some statistical and logical considerations when rescoring tests. *Statistica Sinica, 8*, 713–728.

Gulliksen, H. O. (1950). *Theory of mental tests*. New York: Wiley. (Reprinted, 1987, Hillsdale, NJ: Erlbaum)

Lord, F. M., & Novick, M. R. (1968). *Statistical theories of mental test scores*. Reading, MA: Addison-Wesley.

Ree, M. J., & Earles, J. A. (1991). Predicting training success: Not much more than g. *Personnel Psychology, 44*, 321–332.

Ree, M. J., & Earles, J. A. (1992). Intelligence is the best predictor of job performance. *Current Directions in Psychological Science, 1*, 86–89.

Ree, M. J., & Earles, J. A. (1993). g is to psychology what carbon is to chemistry: A reply to Sternberg and Wagner, McClelland, and Calfee. *Current Directions in Psychological Science, 2*, 11–12.

Thissen, D., & Wainer, H. (Eds.). (2001). *Test scoring*. Hillsdale, NJ: Erlbaum.

Wainer, H., & Kiely, G. (1987). Item clusters and computerized adaptive testing: A case for testlets. *Journal of Educational Measurement, 24*, 185–202.

Wainer, H., & Lichten, W. (2000). The Aptitude-Achievement Function: An aid for allocating educational resources, with an advanced placement example. *Educational Psychology Review, 12*(2), 201–228.

Wainer, H., Dorans, D. J., Eignor, D., Flaugher, R., Green, B. F., Mislevy, R. L., Steinberg, & Thissen, D. (2000). *Computerized adaptive testing: A primer* (2nd ed.). Hillsdale, NJ: Erlbaum.

Wainer, H., Lukele, R., & Thissen, D. (1994). On the relative value of multiple-choice, constructed response, and examinee-selected items on two achievement tests. *Journal of Educational Measurement, 31*, 234–250.

2

True score theory

God created the integers. All else is the work of man.
Leopold Kronecker (1823–1891)[1]

Historically, methods for scoring tests grew from basic notions of having experts evaluate the responses to the questions and assign a score. When possible, precautions were taken to ensure that the judges, to the extent possible, were not biased in their judgments because of other knowledge they had about the examinee. In ancient China, written answers were sometimes rewritten by a professional scribe to ensure that the examinee's identity was hidden. As the structure of tests changed to include a large number of questions, the scoring methods were modified apace. Typically, each question was assigned a certain number of points and an examinee's score was determined by adding together the number of points associated with the questions that were responded to correctly. This notion underlay test scoring for most of its four-thousand-year history. It was formally codified within the last century with the development of what has come to be called "true score theory."

Why do we need a theory for test scoring? The fundamental idea of a test, once ingested, seems simple. Count up the points and whoever has the most wins. And, indeed when tests are used as contests, this may be enough. But as we pointed out in Chapter 1, this is not the only use of a test. Tests as prods don't need much theory either, but tests as measuring instruments do. Because using a test as a measuring instrument is a relatively new application, it is not surprising that a formal theory of tests did not begin to appear until the 20th century. A measuring instrument must yield more than just an outcome; it must

[1] Kronecker quote is given in Bell (1937, p. 477).

Parts of this chapter are drawn from an essay coauthored with Stephen Sireci and Henry Braun (Sireci, Wainer, & Braun, 1998). We are grateful for their permission to extract portions of it here.

also provide some sense of the accuracy of that outcome. Having a measure of accuracy in our hands provides us with the critical tool for improving the measurement, as well as stating the degree of confidence that we have in that measurement. A new instrument with greater accuracy is better. As modern test theory developed, various measures of accuracy evolved, each providing a picture of a particular kind of accuracy. To what extent is the test measuring the construct of interest? This has come to be called "construct validity." Validity is a subtle concept with many sides, and those interested in a deep and wide-ranging discussion are referred to Messick's (1989) seminal essay. A second aspect of accuracy relates to the stability of the scores and has been termed "reliability." A lawyer might characterize these two concepts in terms of evidence. *Validity* is the extent to which the evidence provided by the test score is relevant for the inferences that are to be drawn, and *reliability* is the amount of the evidence. This characterization makes clear what turns out to be a theorem in test theory (Lord & Novick, 1968, p. 72): that reliability forms an upper bound on validity.[2] The remainder of this chapter provides the rudiments of the most basic of statistical models of test scores, *true score theory*.

The fundamental idea of true score theory can be stated in a single simple equation:

$$\text{observed score} = \text{true score} + \text{error} \tag{2.1}$$

This equation explicitly states that the score we observe on a test is composed of two components, the true score and an error term. The term *true score* has a very specific technical meaning. It is not some secret platonic truth that is available only by looking deep into an examinee's soul. Instead, it is just the average score that we would expect to obtain if the examinee retook extraordinarily similar (parallel) forms of the exam many, many times. The *error term* characterizes the difference between what is observed on a particular occasion and that unobserved long-term average. Such errors are considered to be random and hence unrelated to true score; that is, the distribution of the errors is the same regardless of the size of the true score. This definition requires the errors to have an average size of zero.

Repeating this same discussion in mathematical terms yields an analog to Equation (2.1) for examinee i on test form j:

$$x_{ij} = \tau_i + e_{ij}, \tag{2.2}$$

[2] Actually, it is a more restrictive bound, for validity must be less than the square of reliability.

where x_{ij} is the observed score for examinee i on test form j,

τ_i is the true score for examinee i, and

e_{ij} is the random error component for examinee i on test form j.

These quantities have the following properties:

$$\mathrm{E}(x_{ij}) = \tau_i \tag{2.3}$$

$$\mathrm{E}(e_{ij}) = 0 \tag{2.4}$$

$$\mathrm{cov}(\tau_i, e_{ij}) = 0. \tag{2.5}$$

In much of the discussion in the rest of this chapter, it will be important to collect the scores for many examinees and to study the variability among those scores. It is commonly assumed that the scores of any examinee are uncorrelated with those of any other examinee. Utilizing Equation (2.2) and these properties, we can decompose the variance of the observed scores into two orthogonal components, true score variance and error variance, i.e.,

$$\sigma_x^2 = \sigma_\tau^2 + \sigma_e^2. \tag{2.6}$$

Equation (2.6) follows directly from the definitions of true score and error, but provides us with many tools to study test performance. Obviously, we prefer tests whose error variance (σ_e^2) is small relative to observed score variance (σ_x^2). A test with small error variance would measure an examinee's true score more reliably than would one with a large error variance. We can characterize how reliably a test works by the ratio of error variance to observed score variance (σ_e^2/σ_x^2). If this ratio is close to 0, the test is working well – the observed score has very little error in it. If it is close to 1, the test is working poorly – the variation in observed score is mostly just error. When the ratio is rescaled (Equation (2.7)) so that it takes the value 1 when there is no error and 0 when it is all error it is the test's *reliability*.

$$\text{Reliability} = 1 - \left(\frac{\sigma_e^2}{\sigma_x^2}\right) \tag{2.7}$$

This representation of reliability is intuitively appealing, but is not in a useful form because it cannot be directly computed from observed data; although we can observe σ_x^2, we cannot observe σ_e^2. A slightly different conception, using the idea of parallel test forms, yields an observable quantity.

Before deriving this important equation, we need a formal definition of parallel test forms. Specifically, two forms, say form X and form X', are parallel if

$$\mathrm{E}(x) = \mathrm{E}(x') = \tau \quad \text{and} \quad \sigma_x^2 = \sigma_{x'}^2 \tag{2.8}$$

for all subpopulations taking the test, where x and x' are the scores on form X and form X', respectively.

The correlation of one parallel form with another, $\rho_{xx'}$, is

$$\begin{aligned}\rho_{xx'} &= \frac{\text{cov}(x, x')}{\sigma_x \sigma_{x'}} \\ &= \frac{\text{cov}(\tau + e, \tau + e')}{\sigma_x \sigma_{x'}} \\ &= \frac{[\text{cov}(\tau, \tau) + \text{cov}(\tau, e) + \text{cov}(\tau, e') + \text{cov}(e, e')]}{\sigma_x^2 \, \sigma_{x'}^2}.\end{aligned}$$

The last three terms in the numerator are 0, yielding

$$\rho_{xx'} = \frac{\text{cov}(\tau, \tau)}{\sigma_x \sigma_{x'}}.$$

But $\text{cov}(\tau, \tau) = \sigma_\tau^2$, and from definition (2.8), $\sigma_x = \sigma_{x'}$ so that $\sigma_x \sigma_{x'} = \sigma_x^2$. Combining these results, we get

$$\rho_{xx'} = \frac{\sigma_\tau^2}{\sigma_x^2},$$

but since $\sigma_x^2 = \sigma_\tau^2 + \sigma_e^2$ (from (2.6)), we can rewrite this as

$$\rho_{xx'} = 1 - \left(\frac{\sigma_e^2}{\sigma_x^2} \right). \tag{2.9}$$

Using a similar approach, it is possible to show that

$$\rho_{x\tau}^2 = 1 - \left(\frac{\sigma_e^2}{\sigma_x^2} \right)$$

so that

$$\rho_{xx'} = \rho_{x\tau}^2. \tag{2.10}$$

This result is important since $\rho_{xx'}$ is directly estimable from data, whereas $\rho_{x\tau}^2$ is not. How to estimate $\rho_{xx'}$ well is the subject of a great deal of work that we will only touch on here. One obvious way is to construct two parallel forms of a test, give them both to a reasonably large sample of appropriate people, and calculate the correlation between the two scores. That correlation is an estimate of the reliability of the test. But making up a second form of a test that is truly parallel to the first is a lot of work. An easier task is to take a single form, divide it randomly in half, consider each half a parallel form of the other, and correlate the scores obtained on the two halves. For obvious reasons, such a measure of test reliability is called *split-half reliability*. This yields an estimate of reliability for a test similar to but only half as long as the test we actually gave. Some sort of adjustment is required. A second issue that must be resolved before using the split-half reliability operationally is to figure out how to split the test. Certainly, all splits will not yield the same estimate, and we would not want to base our estimate on an unfortunate division. Let us consider each of these issues in turn.

2.1. The Spearman–Brown formula for a test of double length

Suppose we take a test X, containing n items, and break it up into two half tests, say Y and Y', each with $n/2$ items. We can then calculate the correlation that exists between Y and Y' (call it $\rho_{yy'}$) but what we really want to know would have been the correlation between X and a hypothetical parallel form X' ($\rho_{xx'}$). A formula (Equation (2.11)) for estimating this correlation was developed by Spearman (1910) and Brown (1910) and is named in their honor:

$$\rho_{xx'} = \frac{2\,\rho_{yy'}}{1+\rho_{yy'}}. \tag{2.11}$$

A derivation of Equation (2.11) follows directly from the characteristics of parallel tests and is given on pages 83–84 in Lord and Novick (1968).

To get an idea of how this expansion works, consider the two lines in Figure 2.1. The dashed diagonal line indicates the equality of reliability between the original test and one of double length. The curved line shows the estimated reliability of the test of double length. Note that when reliability of the original test is extreme (0 or 1), doubling its length has no effect. The greatest effect occurs in the middle; a test whose reliability is 0.50 when made twice as long attains a reliability of 0.67.

2.2. Which split half? Cronbach's coefficient α

There are many ways that we can split a test of n items in half (assuming that n is an even number). Specifically, there are

$$\frac{1}{2}\binom{n}{n/2} = \frac{n!}{2\,[(n/2)!]^2}$$

different ways that n items can be split in half. With all of these possible ways to divide the test in half, it can be a difficult decision to determine which one we should pick to represent best the reliability of the test. One obvious way around this problem is to calculate all of them and use their mean as our best estimate. But calculating all of these for any test of nontrivial length is a very

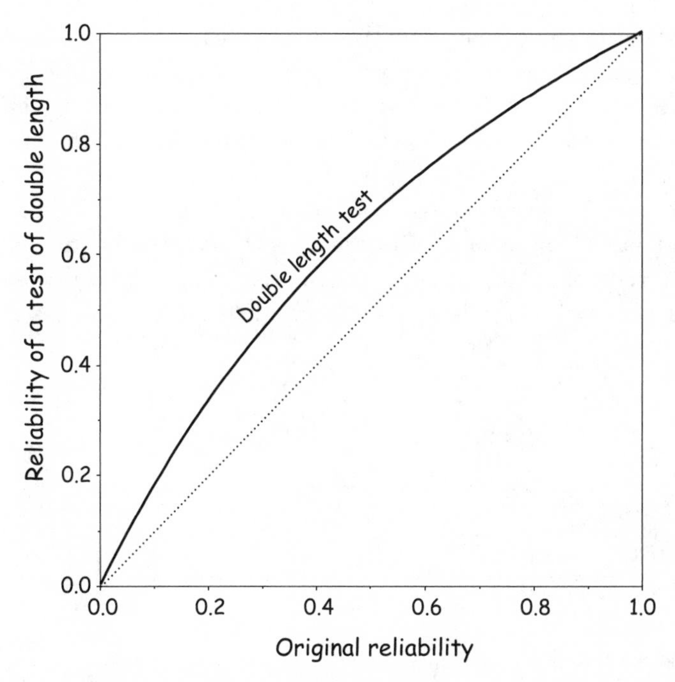

Figure 2.1. Reliability at double length as a function of reliability of a unit length test.

big deal. For example, there are more than 63 trillion (63×10^{12}) split halves for 50 items! One way around this problem is due to Cronbach (1951). He derived a statistic that is a lower bound on the reliability of the test and is equivalent to taking the mean over all possible splits. Mel Novick and Charles Lewis (1967) provided the conditions under which Cronbach's statistic is actually equal to the reliability of the test.[3] Cronbach's statistic is usually called *Cronbach's* α in his honor, and is shown as the expression on the left-hand side of Equation (2.12):

$$\frac{n}{n-1}\left[1-\frac{\sum_{i=1}^{n}\sigma_{y_i}^2}{\sigma_x^2}\right] = \alpha \leq \rho_{xx'}. \tag{2.12}$$

[3] A necessary and sufficient condition for *Cronbach's* α, to be equal to the test's reliability, and not just a lower bound, is that all of the components making up the score have the same true score. The technical name for this condition is that all components be τ-equivalent. The name of this statistic helps to illustrate how unpredictable the path of scientific development can be. Cronbach named this statistics α because he felt that it was but the first in a series of measures of reliability, and the next one (β) would improve upon this one. But before any such improvement could appear, generalizability theory was developed (by Cronbach et al. 1972), which, when combined with the growing popularity of IRT, which uses standard error as a more precise measure of accuracy, reduced the importance of further refinements of a measure of reliability.

To calculate α, we conceive of the test score x as being composed of the sum of n items Y_i. We calculate the variance of each item across all of the examinees who took it and sum it up over all the items. Next, we calculate the variance of the total score. The ratio of these two variances forms the core of *Cronbach's* α.

Of course, *Cronbach's* α is a sample statistic, and, as such, is subject to sample-to-sample fluctuations. One can calculate the standard error of coefficient α for any sample of n items and m examinees using a formula recently provided in Duhachek, Coughlan, and Iacobucci (2005). A 95% bound is given by

$$\hat{\alpha} \pm (1.96)\left[\sqrt{\frac{Q}{m}}\right],$$

where Q is

$$\left[\frac{2n^2}{(n-1)^2(\mathbf{j}'\mathbf{S}\mathbf{j})^3}\right][(\mathbf{j}'\mathbf{S}\mathbf{j})(\operatorname{tr}\mathbf{S}^2+\operatorname{tr}^2\mathbf{S})-2(\operatorname{tr}\mathbf{S})(\mathbf{j}'\,\mathbf{S}^2\,\mathbf{j})] \tag{2.13}$$

and j is $n \times 1$ vector of 1's and S is an $n \times n$ estimated inter-item covariance matrix.

2.3. Estimating true score

Under true score theory, the principal object of interest is the examinee's true score. A reasonable approach is to use the observed score as an estimate of the true score. Although this estimator has the virtue of being unbiased (i.e., equal in expectation to its target), it can be improved upon.

The heuristic behind the improved estimator is similar to the familiar "regression to the mean" argument. Observed scores that fall far below the mean are likely to have resulted, in part, from a relatively larger negative random error on that particular occasion. In a repeat test, we would expect these observed scores to still fall below the mean, but not by nearly so great a margin. The greater the reliability of the test, the less the effect of the random errors.

The improved estimator, due to Kelley (1947), requires three pieces of information: the observed score of the examinee, the mean score over the population of examinees, and a measure of the test's reliability. The estimate $\hat{\tau}$, of the true score τ, is given by

$$\hat{\tau} = \hat{\rho}_{xx'}x + (1-\hat{\rho}_{xx'})\hat{\mu}, \tag{2.14}$$

where $\hat{\rho}_{xx'}$ is an estimate of the reliability of the test score, x is the observed score, and $\hat{\mu}$ is an estimate of the test's population mean score. This result states that to estimate the true score from an observed score, the latter should be regressed toward the mean of the population by an amount related to the test's reliability. The amount of "adjustment" to x increases as the reliability of the test decreases. Note too, that for a fixed level of reliability, the larger the amount of adjustment is, the greater the distance of the observed score from the population mean. We also note that to some readers, Equation (2.14) may be recognizable as a shrinkage estimate more familiar in the Bayesian literature (Stein, 1956).

2.4. A model for error

The reliability coefficient that we have just discussed provides us with one measure of the stability of the measurement. It is often useful to have another measure that can be expressed in the scale of the score. For example, we would usually like to be able to present an estimate of someone's true score with error bounds – $\hat{\tau} = 1.2 \pm 0.3$ with, say, 95% confidence. Such a statement raises two questions. The first is "What does ± 0.3 mean?" The second is "How did you get 0.3?"

There can be different answers to the first question depending on the situation. One common and reasonably useful answer is "it means that 95% of the time that someone whose estimated value of $\hat{\tau}$ is this value, her true score is within 0.3 of it."

The answer to the second question "How did we get it?" requires a little longer explanation. The explanation goes back to Equation (2.9)

$$\rho_{xx'} = 1 - \left(\frac{\sigma_e^2}{\sigma_x^2}\right).$$

Simplifying this equation, we can isolate the error variance on one side of the equation and, by taking the square root, obtain an estimator for the *standard error of measurement*

$$\sigma_e = \sigma_x \sqrt{1 - \rho_{xx'}}. \tag{2.15}$$

The uncertainty in our estimates of true score will be due to the variability of the error (e). To be able to provide a probability statement about the variability of the estimate, we need to assume something about the distribution of the error. What sort of distribution we assume depends on the character of the scoring metric we use for the test. There are many choices (see Lord & Novick, 1968,

Chapter 23, for an extended discussion of alternatives), but for the common situation where observed scores fall in the middle of the range of possible scores, the assumption of a Gaussian distribution of errors is reasonable for most practical applications.

Thus, in conclusion, true score theory provides a wonderful way to summarize the efficacy of a test, but another approach was needed to handle gracefully the richer data structures found when the unit being considered was smaller than the entire test. In Chapter 3, we describe a test scoring approach that focuses on the individual item and allows us to aggregate up from item statistics to characterize the whole test. This next approach is aptly named item response theory.

Questions

1. What is true score theory?
2. What is the most basic equation of true score theory?
3. What is validity in terms of evidence?
4. What is reliability?
5. Can a test be more valid than it is reliable?
6. How is a test's reliability related to its length?

References

Bell, E. T. (1937). *Men of mathematics*. New York: Simon & Schuster.

Brown, W. (1910). Some experimental results in the correlation of mental abilities. *British Journal of Psychology, 3*, 296–322.

Cronbach, L. J. (1951). Coefficient alpha and the internal structure of tests. *Psychometrika, 16*, 297–334.

Cronbach, L. J., Gleser, G. C., Nanda, H., & Rajaratnam, N. (1972). *The dependability of behavioral measurements: Theory of generalizability for scores and profiles*. New York: Wiley.

Duhachek, A., Coughlan, A. T., & Iacobucci, D. (2005). Results on the Standard Error of the Coefficient Alpha Index of Reliability. *Marketing Science, 24*(2), 294–301.

Kelley, T. L. (1947). *Fundamentals of statistics*. Cambridge: Harvard University Press.

Lord, F. M., & Novick, M. R. (1968). *Statistical theories of mental test scores*. Reading, MA: Addison-Wesley.

Messick, S. (1989). Validity. In R. Linn (Ed.), *Educational measurement* (pp. 13–104). Washington, DC: American Council on Education.

Novick, M. R., & Lewis, C. (1967). Coefficient alpha and the reliability of composite measurements. *Psychometrika, 32*, 1–13.

Sireci, S., Wainer, H., & Braun, H. (1998). Psychometrics: An overview. In P. Armitage & T. Colton (Eds.), *Encyclopedia of biostatistics* (pp. 327–351). London: Wiley.

Spearman, C. (1910). Correlation calculated with faulty data. *British Journal of Psychology, 3*, 271–295.

Stein, C. (1956). Inadmissability of the usual estimator for the mean of a multivariate normal distribution. *Proceedings Third Berkeley Symposium on Mathematical Statistics and Probability, 1*, 197–206. Berkeley: University of California Press.

3
Item response theory

Item response theory (IRT) is a family of mathematical descriptions of what happens when an examinee meets an item. It stems from early notions that test items all ought to somehow measure the same thing (Loevinger, 1947). Originally, IRT formalized this by explicitly positing a single dimension of knowledge or underlying trait on which all examinees rely, to some extent, for their correct responses to all the test items (although recent research has generalized this to allow multidimensionality (Reckase, 1985, 1997; McDonald, 2000; Segall, 2001)). Examples of such traits are verbal proficiency, mathematical facility, jumping ability, or spatial memory. The position that each item occupies on this dimension is termed that item's *difficulty* (usually denoted as b); the position of each examinee on this dimension is that examinee's *proficiency* (usually denoted as θ). The IRT model gives the probability of answering a question correctly in terms of the difference between b and θ (both of which are unobservable). The simplest IRT model combines just these two elements within a logistic function.[1] Because it characterizes each item with just a single parameter (difficulty $= b$), in its logistic form it is called the *one parameter logistic model* or 1-PL. This model was first developed and popularized by the Danish mathematician Georg Rasch (1901–1980), and so is often termed the *Rasch Model* in his honor.[2] We shall denote it as the 1-PL model in this account to reinforce its position as a member of a parametric family of logistic models.

[1] The logistic is not the only kind of response function employed for joining ability and difficulty; several different kinds of ogive functions have been used. The cumulative normal is a popular alternative, but even monotone splines have proved to be sometimes useful (Ramsay, 1991, 1992).

[2] There is a subtle, but important difference between the Rasch model and the 1-PL model, although they are in some formal sense identical. Specifically, the Rasch model is as specified in Equation (3.1), and the 1-PL model has a fixed slope a in the logit $[a(\theta - b)]$ that must be estimated from the data. Obviously, the Rasch model sets a equal to 1. The place where this comes home to roost is in the variance of the distribution of θ. In the 1-PL model, one arbitrarily

The 1-PL model is

$$P(\theta) = \frac{1}{1 + e^{-(\theta - b)}}, \tag{3.1}$$

where $P(\theta)$ is the probability of someone with proficiency θ responding correctly to an item of difficulty b. The interpretation of P can be thought about in two quite different ways: (i) arising from sampling examinees and (ii) arising from having probability-in-the-soul of the examinee, what Holland (1990) calls the "stochastic subject" rationale.

3.1. An obiter dictum on *P*

For reasons that we shall get to shortly, we tend to favor the sampling interpretation of P. This approach assumes that when confronted with an item, an examinee either knows the answer or does not. The probability P arises from the idea that if there were a large number of examinees with proficiency θ, P of them will be able to answer that item and $(1 - P)$ will not. The proportion of examinees with proficiency θ who can answer a particular item, with difficulty b, correctly is given by Equation (3.1). $P(\theta)$ is increasing in θ for a fixed b; for a fixed θ, $P(\theta)$ is smaller for larger values of b.

The "probability-in-the-soul" idea stems from quite a different conception. It asserts that there is a probability P for a given person answering a given item correctly. The notion is that if one could observe a response and then wipe clean the examinee's memory and readminister the item over and over again, the examinee would answer it correctly P proportion of the time. This point of view is left over from the psychophysical origins of test theory, in which experimental subjects could be presented with a very weak stimulus (e.g., a dim light, a barely audible sound, a very light touch) and asked if they perceived (e.g., heard, saw, felt) it. Sometimes they would and sometimes they would not. A stimulus strength that was perceived by the subject 50% of the time was called the "threshold strength" of the stimulus. This is directly analogous to the item's difficulty, and so it is easy to see how the conception of building the stochastic variability into the person within test theory was a natural outgrowth of the past.

fixes the variance of the proficiency distribution to set the scale of the model (e.g., $\theta \sim N(0, 1)$ is a common structure). In the Rasch model, setting a equal to 1 fixes the scale and so the variance of θ must be estimated. Which version is to be preferred depends on the circumstance. If two forms of the same test are given to two randomly selected subsamples from a population, it is plausible to assume that the distributions of θ are the same, and so one should estimate a. In addition, this approach generalizes directly to the more complex parameterizations required in operational situations, and so is the one that is usually preferred in practice.

Bush and Mosteller (1955) adopt the probability-in-the-soul approach in their discussion of probabilistic learning models. They argue that this approach is an inescapable fact of life.

> .... that performance is an unpredictable thing, that choices and decisions are an ineradicable feature of intelligent behavior, and that the data we gather from psychological experiments are inescapably statistical in character.
>
> *(p. 336)*

In early work on IRT, some psychometricians also adopted this view. For example, Rasch (1960) explained how to interpret his model as,

> .... certain human acts can be described by a model of chance .... Even if we know a person to be very capable, we cannot be sure he will solve a certain difficult problem, nor even a much easier one. There is always a possibility that he fails – he may be tired or his attention is led astray, or some other excuse may be given. Furthermore if the problem is neither 'too easy' nor 'too difficult' for a certain person the outcome is quite unpredictable. But we may in any case attempt to describe the situation by *ascribing to every person a probability of solving each problem correctly....*
>
> *(p. 73)*

And even Lord and Novick (1968) in their cornerstone of modern test theory take a similar tack (p. 30) when they describe the "transient state of the person" giving rise to a "propensity distribution."

Although probability-in-the-soul may sometimes be a useful way to think about how examinees respond (it is hard to imagine how models such as (3.3), which include guessing, would've evolved without such a conception), most psychometricians believe that the sampling approach is a better description of the phenomenon encountered in testing.

Birnbaum (1968) was one early advocate of this view,

> Item scores ... are related to an ability θ by functions that give the probability of each possible score on an item for a randomly selected examinee of given ability.
>
> *(p. 397)*

And Lord (1974) appeared to join in this view,

> An alternative interpretation is that $P_i(\theta)$ is the probability that item i will be answered correctly by a randomly chosen examinee of ability θ.
>
> *(p. 250)*

but on the same page he leaves room for some probability-in-the-soul as well (in this paper, it is regarding responses to previously omitted items),

> These interpretations tell us nothing about the probability that a specified examinee will answer a specific item correctly.
>
> *(p. 250)*

Thus, there appears to be room for both conceptions. There is, however, a dark side to the probability-in-the-soul approach. By adopting this outlook, you can be fooled into believing that you can ignore the need for ever mentioning the examinee population. Holland (1990) points out that

> ... this is an illusion. Item parameters and subject abilities are always estimated relative to a population, even if this fact may be obscured by the mathematical properties of the models used. Hence I view as unattainable the goal of many psychometric researchers to remove the effect of the examinee population from the analysis of test data.
>
> *(p. 585)*

Thus, the interpretation of P as probability-in-the-soul too easily leads one to the chimera of "sample-free item analysis" (Wright & Douglas, 1977). Again, quoting Holland, "The effect of the population will always be there, the best that we can hope for is that it is small enough to be ignored."

A number of different procedures have been proposed to estimate parameters of IRT models. Lord (1967) provided a computer program to compute unconditional maximum likelihood (UML) and joint maximum likelihood (JML) estimates in the same year that Bock (1967) described a method for estimation by maximizing the marginal likelihood (MML). Rasch (1960) suggested the method of conditional maximum likelihood (CML) estimates whose properties were established by Andersen (1970).

The justification for MML lies squarely on the sampling view of P, and indeed, from this point of view it is *the* maximum likelihood estimate; the term marginal is redundant. CML and UML are both approximations to MML, and the justification for both of them rests squarely on the conception of P being generated internally, without any sampling considerations.

The philosopher Karl Popper (1902–1994) conceived of probability as the same sort of idea as the electron or the quark (Popper, 1987); a mathematical convenience that is valuable only so long as it provides observable, and hence testable, outcomes. He used this idea to get away from the frequentist's often untenable assumption of a sample from an infinitely repeatable process while still distinguishing it from a Bayesian personal probability. Both conceptions of the P generated by IRT models suit Popper's definition, and each may have value in certain circumstances, but the sampling conception is the one that, historically, has led us to reasonable practice.

Those interested in further details of how the various estimation methods rely on the character of the process that is assumed to have generated P are referred to Paul Holland's (1990) presidential address to the Psychometric Society, from which much of this discussion was abstracted.

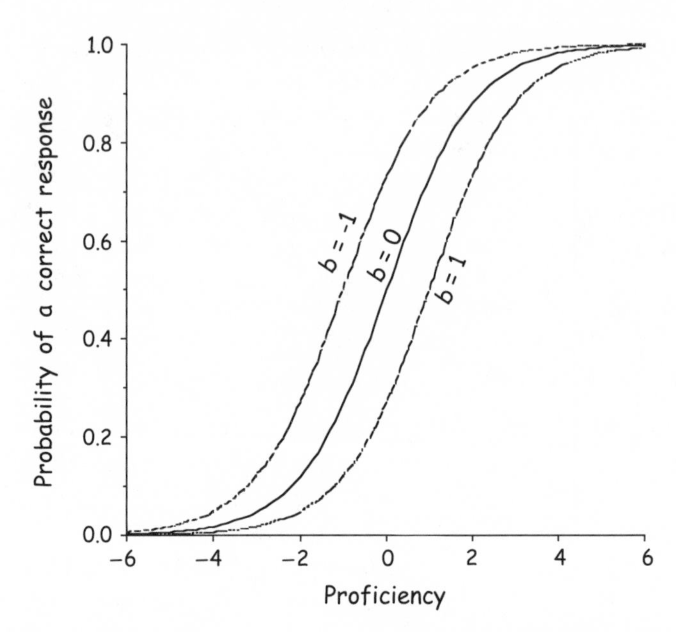

Figure 3.1. Item characteristic curves for the 1-PL model at three levels of difficulty.

With this perhaps overlong aside on the meaning of P completed, let us return to a discussion of the character of the 1-PL model.

3.2. Returning to the 1-PL

The structure of this model is most easily seen in a graph. In Figure 3.1 is a plot of what this function looks like for three items of different difficulty. These curves are called the *item characteristic curves* (or ICCs – they are also sometimes referred to as *trace lines* or *item response functions*). Note that the ICCs for this model are parallel to one another. This is an important feature of the 1-PL model. It is informative to contrast the ICCs shown in Figure 3.1 with those in Figure 3.2.

The 1-PL model has many attractive features, and the interested reader is referred to the writings of Georg Rasch (1960) and of Ben Wright (Wright & Stone, 1979) for convincing descriptions of its efficacy.

In many applications of IRT to predetermined domains of items, it has been found that one does not get a good fit to the data with the 1-PL model. A common cause of misfit is that the ICCs of all items are not always parallel. When this occurs, there are two options open. One is to delete items whose ICCs show slopes that are divergent from the mass of other items. This Procrustean notion

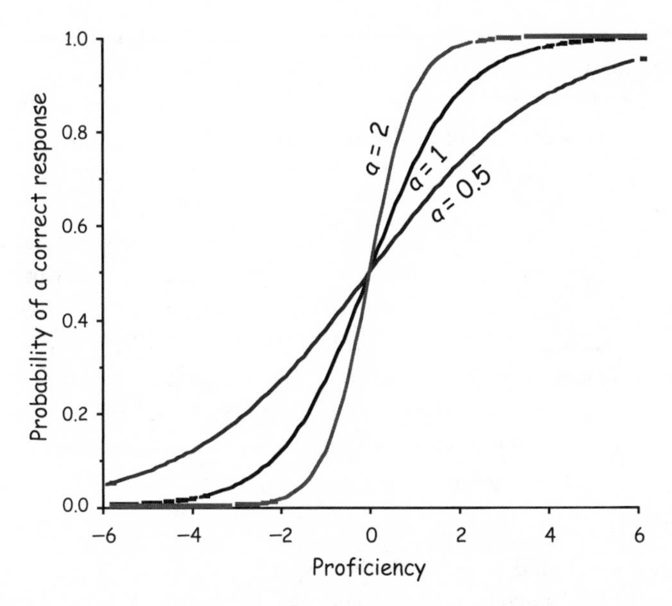

Figure 3.2. Typical item characteristic curves for the 2-PL model, all with the same difficulty.

of test construction is roughly analogous to shooting arrows into a tree and then drawing a target around them. The second is to generalize the model to allow for different slopes. This can be done through the addition of a second parameter for each item. This parameter, usually denoted as a, characterizes the slope of the item characteristic curve, and is often called the *item's discrimination*. The resulting mathematical model, which now contains two parameters per item, is called the 2-PL model and looks quite similar to the 1-PL model. Explicitly it is

$$P(\theta) = \frac{1}{1 + e^{-a(\theta - b)}}. \tag{3.2}$$

Once again our intuition is aided by seeing plots of the ICCs achievable with this more general model. We have drawn three 2-PL ICCs for items with the same b parameter ($b = 0$) in Figure 3.2, demonstrating the variation in slopes often seen in practice. Shown is an item that has rather high discrimination ($a = 2$), another with average discrimination ($a = 1$), and a third with lower than average discrimination ($a = 0.5$).

The reason for calling the maximum value of the slope of the ICC "discrimination" is that items with large slopes (i.e., high discrimination) are better able to distinguish between lower and higher proficiency examinees. For an item of high discrimination, there is a relatively short interval along the proficiency scale in which $P(\theta)$ moves from nearly 0 to nearly 1. Note that items of high discrimination aren't very useful unless they are centered (i.e., have a b value) in a region of the proficiency scale of interest to the examiner.

With the addition of the slope parameter, the 2-PL model greatly expanded the range of applicability of IRT. Many sets of items that could not fit under the strict equal slope assumption of the 1-PL model could be calibrated and scored with this more general model. However, this was not the end of the trail. As long as the multiple-choice item remains popular, the specter of an examinee getting an item correct through guessing remains not only a real possibility but also an event of substantial likelihood. Neither of the two models so far discussed allows for guessing – if an examinee gets an item correct, it is assumed to provide evidence for greater proficiency. Yet, sometimes we see evidence in a response pattern that an examinee has not obtained the correct answer in a plausible fashion. Specifically, if someone gets a very difficult item correct, an item far beyond that examinee's estimated proficiency, there are two possible explanations.

1. The test is not unidimensional – the suspect item was answered using a skill or knowledge base other than the one we thought we were testing. This can be corrected by either modifying the test (removing the offending item) or generalizing the model to allow for multidimensionality. The former fix is not likely to work because different people may choose different items to guess or cheat on – eventually we will have to eliminate all difficult items.[3] The latter fix is feasible, and is in part discussed here as testlet effects can be considered subtest abilities.
2. The other possibility is that the item's answer was guessed. So, if we are to continue to use multiple-choice items, we have little choice but to use a more general model to describe examinees' performance on them. Such a model was fully explicated by Allan Birnbaum in Lord and Novick's (1968) comprehensive text. It adds a third parameter, c, that represents a Bernoulli floor on the probability of getting an item correct. The resulting model, not surprisingly called the *three-parameter logistic model* and denoted 3-PL, is shown explicitly in the following equation:

$$P(\theta) = c + \frac{1-c}{1+e^{-a(\theta-b)}}. \tag{3.3}$$

Once again, we can get a better feel for the structure of the 3-PL model after we view a plot of its typical ICC. Such a plot is shown in Figure 3.3.

There is indeterminacy in the estimation of the parameters of all of these models that must be resolved one way or another. For example, suppose we

[3] In addition, as items age their security becomes more problematic, hence older items, or items that have been used more broadly, or more frequently, may need to be treated differently.

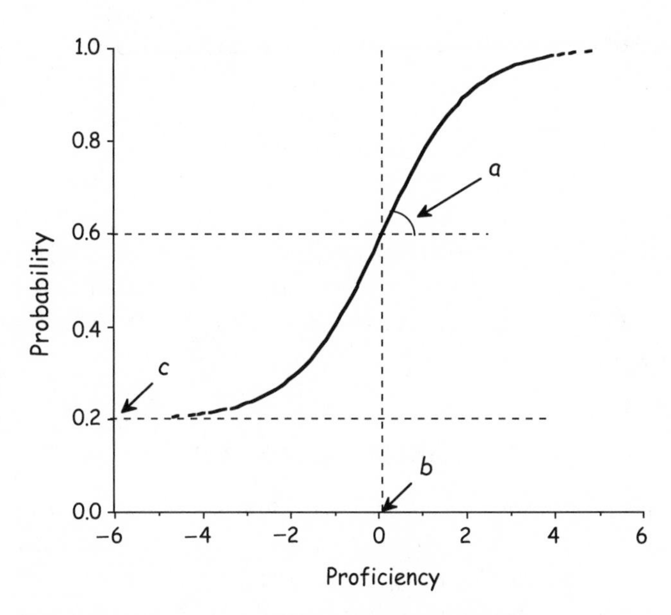

Figure 3.3. Typical item characteristic curve for the 3-PL model.

define a new value of the slope, say a^* as $a^* = a/A$, where a is the original value of the slope and A is some nonzero number. If we then define a new difficulty as $b^* = Ab + B$, where b is the original difficulty and B is some constant, a redefinition of proficiency as $\theta^* = A\theta + B$ would yield the result that $P(\theta, a, b, c) = P(\theta^*, a^*, b^*, c)$. Obviously, there is no way to tell which set of parameters is better because they produce identical estimates of the probabilities of correct responses, and hence provide exactly the same fit to the observed data. The usual way of resolving this indeterminacy is to scale proficiency so that θ has a mean of 0 and a standard deviation of 1 in some reference population of examinees. This standardization allows us to understand at a glance the structure of our results. However, if we separately standardized with respect to two independent samples that were not randomly drawn from the same population, we cannot directly compare the resulting item parameter estimates between them (because the first two moments of their proficiency distributions have been made identical artifactually). This unfortunate practice too often occurs in international comparisons when two forms of a test given in different languages are scaled separately. Comparisons between the two groups are then made as if the scales were identical. To do this properly, we need to link these independent samples. A method for doing this is described in Wainer and Mislevy (2000).

The 3-PL model is the IRT model that is most commonly applied in large-scale testing applications, and so we shall confine the balance of our discussion to it. Estimating the parameters for a set of items under an IRT model is usually called calibration. As indicated earlier, this is a difficult task since the examinees' proficiencies (θ values) must be estimated as well. Some widely used programs available for maximum likelihood calibration are BILOG for dichotomous data (Mislevy & Bock, 1983) and MULTILOG for polytomous data (Thissen, 1991). In this book, our primary focus is on a Bayesian approach to model characterization and parameter estimation; however, for purposes of historical completeness and to provide an informed comparison, we will first provide a brief description of the most widely accepted method of maximum likelihood estimation.

3.3. Estimating proficiency

In this section, we assume that we have already estimated (somehow) the three parameters (a, b, and c) for each item, and have given this calibrated test to a sample of examinees. Our task is to estimate the proficiency (θ) for each of them. We do this by using the method of maximum likelihood estimation. In this discussion, we alternate between the method of maximum likelihood and Bayes modal estimates. In the way we use these terms, there is a clear connection between the two, because the *maximum likelihood estimator is conceptually the same as a Bayes modal estimator with an improper uniform prior on proficiency.*

To estimate proficiency, we need to define three new symbols:

x_i is the vector of item responses for examinee i, in which each response is coded 1 if correct, and 0 otherwise; it has elements $\{x_{ij}\}$, where the items are indexed by j.

β_j is the item parameter vector (a_j, b_j, c_j) for item j and is a vector component of the matrix of all item parameters β and $Q(\theta) = 1 - P(\theta)$.

The conditional probability of x_i given θ and β (the likelihood) is shown in the following equation:

$$P(x_i|\theta_i, \beta) = \prod_j P_j(\theta_i)^{x_{ij}} Q_j(\theta_i)^{1-x_{ij}}. \tag{3.4}$$

The genesis of Equation (3.4) should be obvious with a little reflection. It is merely the product of the model-generated probabilities (the ICCs) for each item. The first term ($P(\theta)$) in the equation reflects the ICC for correct responses (when $x_{ij} = 1$); the second term ($Q(\theta)$), for incorrect responses (when $x_{ij} = 0$). Sometimes this is better understood graphically. Suppose we consider a two-item test in which an examinee gets the first item correct and the second item incorrect. The probabilities for each of these occurrences as a function of θ are shown in the top and middle panels of Figure 3.4, respectively.

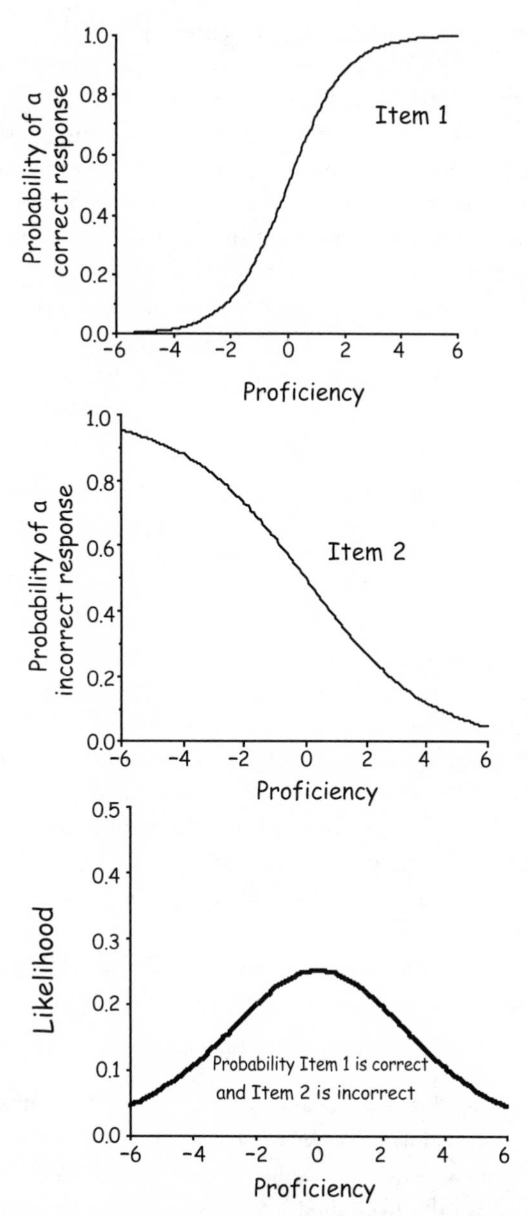

Figure 3.4. A graphical depiction of how item trace lines multiply together to yield the posterior likelihood. This is a graphic version of Equation (3.4).

In order for this product to represent validly the probability of a particular response vector, the model must be a close approximation of the process that generated the data, and the item responses must be conditionally independent. Conditional independence is a basic assumption of most IRT models. It means that the probability of answering a particular item correctly is independent of responses to any of the other items once we have conditioned on proficiency (θ). This assumption is testable (see Rosenbaum, 1988), and, when it is violated, tends to yield overestimates of the accuracy of estimation (Thissen, Steinberg, & Mooney, 1989). The balance of this book, from Chapters 4 onwards, provides expanded test scoring models to allow us to deal properly with situations when simple conditional independence given θ is not a credible assumption.

3.4. Example of how ICCs multiply to yield the likelihood of proficiency

If we know β, the item parameters, we can look upon Equation (3.4), for a fixed response pattern x_i, as the likelihood function $L(\theta|x_i)$ of θ given x_i, its value at any value of θ indicates the relative likelihood that x_i would be observed if θ were the true value. Equation (3.4) thus conveys the information about θ contained in the data and serves as a basis for estimating θ by means of maximum likelihood or Bayesian procedures. The *maximum likelihood estimate* of θ is merely the mode of the likelihood. Stated graphically, in terms of Figure 3.4, it is the value of θ associated with the highest point on the likelihood (the bottom panel in the figure). This likelihood was obtained by multiplying the curve in the top panel by the curve in the middle panel. The estimation methods commonly used are variations on this theme.

Another common method for estimating proficiency is the *Bayes modal estimate*. It is based on the posterior distribution

$$P(\theta|x_i) \propto L(\theta|x_i)p(\theta), \tag{3.5}$$

where $p(\theta)$ expresses knowledge about θ prior to the observation of x_i. Thus, this is commonly called the *prior distribution of* θ. To accommodate the prior distribution into the estimation scheme, we merely treat it as one more item and multiply it with the likelihood, along with everything else. If we want to reflect no prior information whatsoever, then our choice must be to insist that $p(\theta)$ has the same value for all θ – a "uniform prior" – and the posterior distribution for θ is simply proportional to the likelihood function. Alternatively, an "informative prior" has a more profound effect; this is shown graphically in Figure 3.5. The prior used in this illustration is a standard normal distribution,

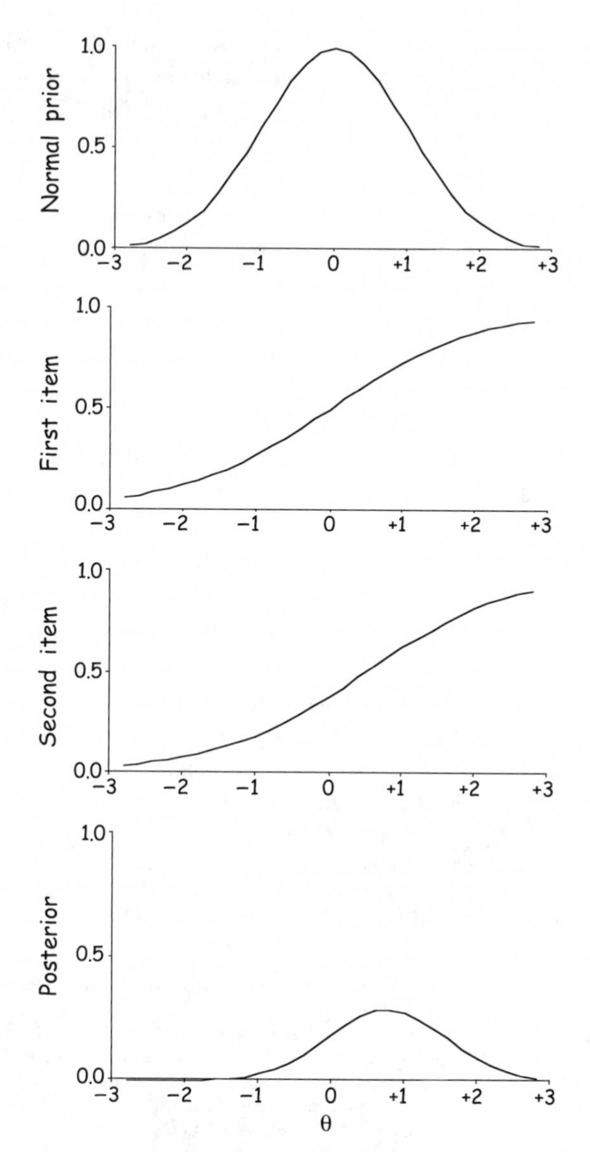

Figure 3.5. Schematic representation of Bayes modal estimation showing a normal prior, responses to two items, and the resulting posterior.

and is shown in the top panel. Note that in this example (taken from Thissen, Steinberg, & Wainer, 1988), the examinee has taken a two-item test, and has gotten both correct. The posterior is shown. If we simply multiply these two ICCs together, to get the probability of occurrence of both events (their likelihood), we would not obtain a curve such as that shown at the bottom of Figure 3.4. Instead, we would get an ogive function that would not allow us to obtain a finite estimate of proficiency. Thus, even though this curve is an important component of the *posterior distribution* of proficiency, it becomes practical only when multiplied by the *prior distribution* of proficiency. In this way, the posterior shows the likelihood of various values of proficiency after (posterior to) the examinee responds. The bottom panel of Figure 3.5 shows this posterior distribution. This representation emphasizes Samejima's (1969) observation that the prior functions as just one more item. In this instance, we can interpret the prior as the correct answer to the question, "Are you part of the population whose proficiency distribution is $N(0, 1)$?"[4]

Although finding the value of θ that maximizes Equation (3.4) cannot be done in closed form, it is in principle straightforward using an iterative method such as Newton–Raphson. In practice, many problems can arise. For example, if an examinee answers all items correctly (or all incorrectly), the estimate of proficiency will not be finite. In the 3-PL model, there are a number of other response patterns (e.g., many patterns at below chance levels) that would also yield infinite proficiency estimates. Also, the likelihood surface does not always yield a single mode; sometimes it can have a number of local extrema. In these cases, solutions to the derivative equations may correspond to a local, but not the global, maximum of l, or even to a local minimum. The problems of infinite estimates are usually solved by utilizing a prior proficiency distribution (that is, using the kind of Bayesian estimator discussed next); those

[4] The inclusion of a prior in the estimation of proficiency was an important breakthrough. Until this became common practice, users of IRT had no principled way to deal with situations in which the examinee got all of the items right (or all wrong). Instead, ad hoc rules were offered (Wright & Stone, 1979, p. 61) in which users were instructed to "edit the binary data matrix of person-by-item responses such that no person has a zero or a perfect score and no item has a zero or a perfect score." Such advice renders the model useless on criterion referenced tests in which it is good news when all examinees get an item correct. It also means that examinees with extreme abilities are not provided scores. Later work of the same genre (Linacre, 2004, p. 244) yields only a slight improvement, suggesting that extreme scores are still dropped, and only later, apparently as an afterthought, suggests that reasonable extreme measures should be imputed for them (using a Bayesian approach) so that all persons and items have measures. The inclusion of a prior provides a principled and graceful solution to this problem. Of course, there remains some arbitrariness, since it is true that if an examinee gets all of the items correct the only thing we know for sure is that her proficiency is likely greater than the most difficult items. But we are pretty much dead certain that her proficiency is not infinite. The inclusion of a prior provides a high, but finite estimate. A parallel story follows for examinees who get all items wrong. The remaining mystery is why some continue using the old approach – Einstein suggested a reason when he quipped that "old theories never die, just the people who believe them." The use of priors is sometimes criticized as adding subjectivity to a supposedly objective score. To some extent, this is true, but some sort of judgment is required to give finite estimates when all responses are the same. As we have just explained, a plausible prior is both principled and public. In addition, experience has shown that this approach typically yields sensible results.

of local extrema are often resolved through the use of a "good" (i.e., close to the global maximum) starting value for the Newton–Raphson iterations (e.g., one based on a rescaled logit of percent-correct), or running mode-finding algorithms from multiple starting values.

3.5. On the accuracy of the proficiency estimate

So far we have been concerned with obtaining a point estimate of an examinee's proficiency. The point we have adopted is the most likely value, the mode of the posterior distribution. But even a quick glance at the posterior in Figure 3.5 tells us that there is a substantial likelihood of other values. The standard deviation of the posterior distribution is commonly used to characterize the precision of the proficiency estimate. If the posterior is very narrow, then we are quite sure that the proficiency estimate we provide is a good one. If the posterior is broad, we are less sure. In practice, we can increase the accuracy of the estimate of proficiency by increasing the length of the test using *appropriately difficult* items. If we add an item that is much too easy for the examinees in question, the ICC would essentially be a horizontal line in the neighborhood of these modal θ's. Multiplying the posterior by a constant such as this would do little to shrink the variance of the posterior. An identical result occurs if the item is much too difficult. This is meant to emphasize the side condition of adding *appropriately difficult* items.

If the number of items administered to an examinee is large, the variance of the posterior estimate function can be approximated as the negative reciprocal of the Fisher *information function*:

$$I(\theta) = \sum_j \frac{(P_j')^2}{P_j(\theta)Q_j(\theta)}, \tag{3.6}$$

where P_j' is the first derivative of P_j with respect to θ. This expression has the attractive features of (a) being additive over items and (b) not depending on the values of the item responses. This means that for any given θ, one could calculate the contribution of information – and therefore to the precision of estimation of θ – from any item in an item pool. A typical information function is shown in Figure 3.6. This approximation for the estimation of error of the maximum likelihood estimate of θ is less accurate for a small number of items, but its advantages make it a popular and reasonable choice for practical work.

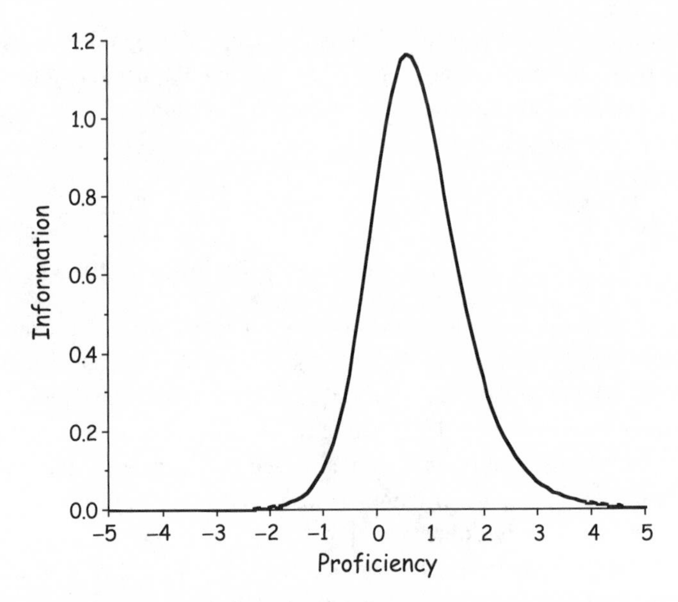

Figure 3.6. Typical 3-PL information function.

A similarly motivated expression can be used to indicate the precision of the Bayes modal estimate of θ:

$$\mathrm{var}^{-1}(\theta|x_t.) \approx I(\theta) - \frac{\partial^2 p(\theta)}{\partial\theta^2},$$

where $p(\theta) = \log P(\theta)$.

The precision of the Bayes modal estimate θ typically exceeds that of the maximum likelihood estimate, because the information from the item responses is augmented by a term that depends on the prior distribution. If $P(\theta)$ is normal with mean μ and variance σ^2, for example, then even before the first item is presented, one has at least this much knowledge about what an examinee's proficiency might be. In this case, $I(\theta) = 0$, since no items have been administered, but $\partial^2 p(\theta)/\partial\theta^2 = -\sigma^{-2}$. The impact of this prior information however, decreases as more items are given because its contribution remains fixed while $I(\theta)$ increases with each item administered. Also note that if the prior is uniform, it contributes nothing to measurement precision.

3.6. Estimating item parameters

Up to this point, we have assumed that the item parameters were known. This is never the case in practice when estimates must be used. Precise estimates with known properties are obviously desirable. In this section, we outline a general framework for item calibration, briefly discuss the pros and cons of a variety of item calibration procedures that have been used in the past, and describe, in some detail, a Bayesian variation of the method of maximum marginal likelihood.

3.6.1. Notation and general principles

The probability of observing the response matrix $X = (x_1, \ldots, x_N)$ from a sample of N independently responding examinees can be represented as

$$P(\mathrm{X}|\theta, \beta) = \prod_i P(x_i|\theta_i, \beta) = \prod_i \prod_j P(x_{ij}\,|\theta_i, \beta_j), \tag{3.7}$$

where $\theta = (\theta_1, \ldots, \theta_N)$ and $\beta = (a_1, b_1, c_1, \ldots, a_J, b_J, c_J)$ are all considered unknown, fixed parameters. The continued product over items for each examinee is understood to run over only those items administered to that examinee, assuming that the responses to the items not administered could be considered ignorable in the sense of Little and Rubin (1987). (See Chapter 12 for an extended discussion of missing data within IRT.)

As we mentioned previously, there are three commonly used approaches to item parameter estimation: *joint maximum likelihood* (JML), which maximizes the likelihood depicted in Equation (3.7); *conditional maximum likelihood* (CML); and *maximum marginal likelihood* (MML). The last is the method of choice. CML is possible only for the 1-PL model, and even there is so computationally intensive as to be impractical in many situations. It is never used for estimation in the 3-PL model, and we do not dwell on it further. The interested reader is referred to Wainer, Morgan, and Gustafsson (1980) for details on this procedure. Before we go on to describe MML, we briefly discuss JML.

JML estimates are obtained by finding the values of each β_j and each θ_i that together maximize Equation (3.7). This is done by applying exactly the same ideas discussed earlier in the context of estimating θ. Direct maximization of Equation (3.7) jointly with respect to θ and β often proves unsatisfactory for a number of reasons. First, for a fixed number of items, JML estimates of β are not consistent in the number of examinees; that is, the expected values of these estimates do not converge to their true values. And, because each of the many θ values is poorly determined when examinees take relatively few items, numerical instabilities can result. Maximizing values of item parameters may thus yield results that are unreasonable or even infinite.

3.6.2. Maximum marginal likelihood estimation

Remedies leading to finite and reasonable estimates of θ under the 3-PL model require information beyond that contained in the item responses, and structure beyond that implied by the IRT model. Let us now discuss a Bayes modal solution, extending ideas introduced in connection with the Bayes modal estimation of θ. Prior distributions for both examinee and item parameters are required. It is perhaps best to develop the solution in two stages.

Let $p(\theta)$ represent prior knowledge about the examinee distribution, assuming we have no additional information to lead us to different beliefs for different examinees. We shall treat $p(\theta)$ as known a priori, but its parameters can be estimated from previous data, or even from the same data as those from which the item parameters are to be obtained. Consistency and increased stability follow if maximizing values for β are obtained after marginalization with respect to $p(\theta)$. That is, MML estimates of β maximize

$$L(\beta|\mathrm{X}) = \int P(\theta, \beta|\mathrm{X})\mathrm{dF}(\theta) \tag{3.8}$$

or, more expansively,

$$L(\beta|\mathrm{X}) = \prod_i \int p(x_i|\theta, \beta)p(\theta)d\theta.$$

Numerical procedures for accomplishing MML estimation are described by Bock and Aitkin (1981), Levine (1985), and Samejima (1983). Without further precautions, however, neither reasonable nor finite item parameter estimates are guaranteed.

Let $p(\beta)$ represent prior knowledge about the item parameter distribution, again assuming for the moment that we have no additional information that leads us to hold different expectations among them. We obtain a posterior distribution for item parameters by multiplying $L(\beta|\mathrm{X})$ by $p(\beta)$:

$$p(\beta|\mathrm{X}) \propto L(\beta\,|\mathrm{X})p(\beta). \tag{3.9}$$

Bayes modal estimates of β are the values that maximize Equation (3.9) (Mislevy, 1986). If a proper distribution $p(\theta)$ has been employed for examinees, and a proper and reasonable distribution $p(\beta)$ has been employed for items, the resulting Bayes modal estimates of β would be stable, reasonable, and consistent. Posterior means can also be employed, but modes are more often used because of their ease of calculation. When estimating item parameters, just as when estimating proficiency, one typically gets indices of the precision with which they have been estimated. Under ordinary MML estimation, one gets standard errors of item parameter estimates; under Bayesian procedures, one gets posterior standard deviations.

The assumption of exchangeability among items (i.e., using the same prior distribution for all items) can be relaxed if some items have been administered previously. The prior distributions for these items can then be determined by the results of previous estimation procedures, taking the forms of distributions concentrated around previous point estimates.

Simulation studies have always been a prerequisite to using any estimation scheme. Asymptotic properties such as consistency do not necessarily characterize estimators' behavior in samples of the size and nature encountered in many specific applications. Moreover, although it is sometimes possible to obtain satisfactory parameter estimates with any of the procedures mentioned (see Mislevy & Stocking, 1987, for a comparison of two methods), the accumulation of evidence suggests that MML, with suitably chosen priors, is the best method, of its type, currently available to obtain item parameter estimates. In the next chapter, we discuss the rationale for broadening the focus of test theory from the item as the fungible unit of test construction to something larger, the testlet.

Questions

1. What is item response theory trying to model?
2. What are the two conceptions of P?
3. What were the historical origins of the probability-in-the-soul conception of P?
4. What are a, b, c, and θ?
5. What is an item characteristic curve?
6. How is a prior like an item?
7. What conception of P yields maximum marginal likelihood as the natural estimation scheme?

References

Andersen, E. B. (1970). Asymptotic properties of conditional maximum likelihood estimators. *Journal of the Royal Statistical Society, Series B, 32*, 283–301.

Birnbaum, A. (1968). Some latent trait models and their use in inferring an examinee's ability. In F. M. Lord & M. R. Novick (Eds.), *Statistical theories of mental test scores* (chaps. 17–20, pp. 397–479). Reading, MA: Addison-Wesley.

Bock, R. D. (1967, March). *Fitting a response model for n dichotomous items*. Paper presented at the annual meeting of the Psychometric Society, Madison, WI.

Bock, R. D., & Aitkin, M. (1981). Marginal maximum likelihood estimation of item parameters: An application of an EM algorithm. *Psychometrika, 46*, 443–459.

Bush, R. R., & Mosteller, F. (1955). *Stochastic models for learning*. New York: Wiley.

Holland, P. W. (1990). On the sampling theory foundations of item response theory models. *Psychometrika, 55*, 577–601.

Levine, M. (1985). The trait in latent trait theory. In D. J. Weiss (Ed.), *Proceedings of the 1982 Item Response Theory and Computerized Adaptive Testing Conference* (pp. 41–65). Minneapolis, MN: Computerized Adaptive Testing Laboratory, Department of Psychology, University of Minnesota.

Linacre, J. M. (2004). A user's guide and manual to WINSTEPS: Rasch-model computer programs. Chicago: Linacre.

Little, R. J. A., & Rubin, D. B. (1987). *Statistical analysis with missing data.* New York: Wiley.

Loevinger, J. (1947). *A systematic approach to the construction and evaluation of tests of ability* (Psychological Monographs, *61*(4), 1–49.

Lord, F. M. (1967). *An analysis of the Verbal Scholastic Aptitude Test using Birnbaum's three-parameter logistic model* (ETS Research Bulletin RB-67-34). Princeton, NJ: Educational Testing Service.

Lord, F. M. (1974). Estimation of latent ability and item parameters when there are omitted responses. *Psychometrika, 39*, 247–264.

Lord, F. M. , & Novick, M. R. (1968). *Statistical theories of mental test scores.* Reading, MA: Addison-Wesley.

McDonald, R. P. (2000). A basis for multidimensional item response theory. *Applied Psychological Measurement, 24*, 99–114.

Mislevy, R. J. (1986). Bayes modal estimation in item response models. *Psychometrika, 51*, 177–195.

Mislevy, R. J., & Bock, R. D. (1983). BILOG: Item and test scoring with binary logistic models [Computer program]. Mooresville, IN: Scientific Software.

Mislevy, R. J., & Stocking, M. L. (1987). *A consumer's guide to LOGIST and BILOG* (ETS Research Report 87-43). Princeton, NJ: Educational Testing Service.

Popper, K. (1987). *The logic of scientific discovery.* London: Hutchinson Publishing Company Ltd. (12th printing).

Ramsay, J. O. (1991). Kernel smoothing approaches to nonparametric item characteristic curve estimation. *Psychometrika, 56*, 611–630.

Ramsay, J. O. (1992). *TESTGRAF: A program for the graphical analysis of multiple choice test and questionnaire data.* Technical report. Montreal, Quebec, Canada: McGill University.

Rasch, G. (1960). *Probabilistic models for some intelligence and attainment tests.* Copenhagen: Denmarks Paedagogiske Institut.

Reckase, M. D. (1985). The difficulty of test items that measure more than one ability. *Applied Psychological Measurement, 9*, 401–412.

Reckase, M. D. (1997). The past and future of multidimensional item response theory. *Applied Psychological Measurement, 21*, 25–36.

Rosenbaum, P. R. (1988). A note on item bundles. *Psychometrika, 53*, 349–360.

Samejima, F. (1969). *Estimation of latent ability using a response pattern of graded scores* (Psychometrika Monographs Whole No. 17). Richmond, VA: Psychometric Society.

Samejima, F. (1983). Some methods and approaches of estimating the operating characteristics of discrete item responses. In H. Wainer & S. Messick (Eds.), *Principals of modern psychological measurement* (pp. 159–182). Hillsdale, NJ: Erlbaum.

Segall, D. O. (2001). General ability measurement: An application of multidimensional item response theory. *Psychometrika, 66*, 79–97.

Thissen, D. (1991). MULTILOG user's guide (Version 6). Mooresville, IN: Scientific Software.

Thissen, D., Steinberg, L., & Mooney, J. A. (1989). Trace lines for testlets: A use of multiple-categorical-response models. *Journal of Educational Measurement 26*, 247–260.

Thissen, D., Steinberg, L., & Wainer, H. (1988). Use of item response theory in the study of group differences in tracelines. In H. Wainer & H. Braun (Eds.), *Test validity* (pp. 147–169). Hillsdale, NJ: Erlbaum.

Wainer, H., & Mislevy, R. J. (2000). Item response theory, item calibration and proficiency estimation. In H. Wainer, D. J. Dorans, R. Flaugher, B. F. Green, R. J. Mislevy, L. Steinberg, & D. Thissen (Eds.), *Computerized adaptive testing: A primer* (2nd ed., chap. 4, pp. 65–102). Hillsdale, NJ: Erlbaum.

Wainer, H., Morgan, A., & Gustafsson, J.-E. (1980). A review of estimation procedures for the Rasch model with an eye toward longish tests. *Journal of Educational Statistics, 5*, 35–64.

Wright, B. D., & Douglas, G. A. (1977). Best procedures for sample-free item analysis. *Applied Psychological Measurement, 14*, 281–295.

Wright, B. D., & Stone, M. H. (1979). *Best test design*. Chicago: MESA Press.

4

What's a testlet and why do we need them?

4.1. Introduction

In 1987, Wainer and Kiely proposed a name for a packet of test items that are administered together; they called such an aggregation a "testlet." Testlets had been in existence for a long time prior to 1987, albeit without this euphonious appellation. They had typically been used to boost testing efficiency in situations that examined an individual's ability to understand some sort of stimulus, for example, a reading passage, an information graph, a musical passage, or a table of numbers. In such situations, a substantial amount of examinee time is spent in processing the stimulus, and it was found to be wasteful of that effort to ask just one question about it. Consequently, large stimuli were typically paired with a set of questions. Experience helped to guide the number of questions that were used to form the testlet. It is easy to understand that if, for example, we were to ask some questions about a 250-word-reading passage, we would find that as we wrote questions, it would get increasingly difficult to ask about something new. Thus, we would find that eventually the law of diminishing returns would set in and a new question would not be generating enough independent information about the examinee's ability to justify asking it. In more technical language, we might say that the within-testlet dependence among items limits the information that is available from that 250-word passage.

Thus, for the first century or so of its existence, testlets were used simply as a method of constructing tests that contained large stimuli in an efficient way. This all changed in the 1980s when computerized adaptive testing (CAT) became technically and economically possible. A CAT is a test that is tailored to the demonstrated ability of the examinee. It is well known (Lord & Novick, 1968) that a test item provides the most information about an examinee when the difficulty of the item is the same as the examinee's ability. This is a self-evident truth. Suppose we are trying to determine an athlete's high-jumping ability by asking her to jump over a sequence of hurdles. If the first hurdle is 20-feet tall, we learn very little when the athlete fails to clear it. Similarly, if the first hurdle is 6-inches high, we learn nothing when it is successfully cleared. If we wanted to ascertain the athlete's ability to an accuracy of, say, 1 inch, we might begin the test with a 6-inch hurdle and continue with 140 hurdles, one every inch, until, say, 12 feet. From the pattern of standing and fallen hurdles, we could estimate the athlete's ability. But it would take 138 hurdles to do it. Suppose the athlete could jump over a 6-foot hurdle, but would miss one at 6 feet 1 inch. We could tell because the first 66 hurdles would be cleared (assuming perfect performance) and the remaining 72 would be knocked over. Note that this "test" would be equally accurate for any athlete whose jumping ability lies between 6 inches and 12 feet. But it takes 138 hurdles to obtain this accuracy, and most of the hurdles are redundant and hence yield little or no information. Now suppose we have an adaptive test. We might use a 6-foot hurdle to begin with. If it is successfully cleared, we might use a 9-foot one (halfway between 6 and 12 feet). If that is missed, we then use a 7.5-foot one (halfway between 9 and 6 feet). If that were missed we would use a 6.75-foot jump etc. In this way, we could measure the performance of anyone whose ability lies in this wide range to within about 1 inch on a test of less than 10 items. That is the power of an adaptive test.

Of course on most tests people don't perform without error. To continue with the jumping example, we might find someone who would clear 6 feet 1 inch after missing at 6 feet, and so a sensible item selection algorithm (the way we choose what is to be the next hurdle (the next item)) should not be quite as drastic as the "split the difference" method described previously. Practical, item selection algorithms tend to move in smaller increments, depending on the character of the item pool and the pattern of the examinee's responses. If the pool is rich in items, having a broad choice at every level of difficulty, the algorithm can pick an item at or near the level desired. If it is sparser, the gaps between items may be suboptimal. So, for example, if a hurdler's true ability is 2.5 feet and after she clears a 2-foot hurdle, the only taller hurdles we have are 3 feet and above, we can never estimate her ability with an error of less than 6 inches. This lack of precision is due to limitations of the item pool.

If the examinee responds in a regular way, getting difficult items wrong and easy items correct, convergence to an accurate estimate of proficiency can be done quickly. If the response pattern is more ragged, convergence is slower and the resulting estimate is less precise. Returning to our overworked hurdler, if she knocks over a 2-foot hurdle, but clears a 30-inch one, then misses at

34 inches but clears a 36-inch one before missing all the rest, our estimate is inaccurate due to the hurdler's inconsistency, not a shortcoming of the item pool.

This example illustrates how the precision of a test can be affected by both the character of the test and the character of the examinee. The first can be controlled by the test developer, the second cannot.

The promise of adaptive testing has enthralled psychometricians for more than thirty years. It began with Lord's initial discussion of his "flexilevel test" in 1971 and, in the same year, his use of the Robbins–Monro procedures to automate it. Lord's later work (1971a, 1971b, 1974, 1980) extended and deepened the initial ideas. David Weiss's preliminary experience in military testing (1974, 1982; Weiss & Kingsbury, 1984) showed how CAT could be practical. Finally, a broad consilience was provided in the enthusiastic "how-to" manual by Wainer et al. in 1990. But in the decade following the publication of this *CAT Primer,* adaptive tests were implemented in some very large operational programs: the U.S. military's ASVAB, the Graduate Management's GMAT, the Graduate Record Examination (GRE), and many others, yielding millions of adaptive test administrations a year. The data thus generated have provided illumination on issues that previously were cloaked in darkness. We discovered that now that we had a pretty clear understanding of how to do adaptive testing, we needed to focus much more seriously on when – and more specifically, on when not to. Chapter 10 of the 2nd edition of the *CAT Primer* (Wainer et al., 2000) provides a more clear-eyed look at these issues than did its predecessor.

One of the issues that emerged from all of this experience relates to a fundamental assumption underlying CAT – that an item's characteristics remain constant regardless of its context – this is usually called the assumption of item fungibility (these are also known as context effects in survey design). This assumption within our jumping test would mean that a 3-foot hurdle is as difficult to jump over whether it follows a 1-foot hurdle or a 6-foot one; obviously a crucial assumption if an item-selection algorithm is to function properly. It is also an assumption whose credibility is rarely tested in operational CATs.

This is but one example of the kind of issues that arise when tests are custom-built for each examinee, one item at a time. In the balance of this chapter, we will discuss a number of such problems in greater detail and indicate what have been the traditional approaches toward solving them. Then we shall conclude with a discussion of how increasing the size of the fungible unit of test construction to the testlet ameliorates these problems.

4.2. Problems

4.2.1. Context effects

What is a context effect?

The term *context effect* refers to any influence that an item may acquire purely as a result of its relationship to the other items making up a specific test. In a traditional testing situation in which everyone takes the same test, context effects, such as they exist, will be the same for all examinees. And so, although they still may induce interpretative errors, there is less concern about their differential impact.

In the case of CAT, there may be a much greater problem of differential context effects, because every examinee potentially takes a different test. Thus, each item's context is potentially different for each examinee. Furthermore, in the CAT real-time approach to test construction, the phase of test development where context effects might ordinarily be identified and eliminated from nonadaptive tests has no counterpart.

Item location

One undesirable item context effect is the influence that item location within a test, or in different tests, has on the item parameter estimates. Several studies (Eignor & Cook, 1983; Kingston & Dorans, 1984; Whitely & Dawis, 1976; Yen, 1980) have found differences in difficulty parameter estimates for items as a function of their location. Whitely and Dawis found that when 15 analogy items were embedded in seven different unspeeded 60-item tests, and parameters were then obtained separately for each test, the estimates differed significantly for 9 of the 15 items. Yen found difficulty differences due to context effects for reading comprehension items administered in a unspeeded test, with the items becoming more difficult the later in the test they appeared. A preequating study conducted by Eignor and Cook replicated Yen's findings.

In a study on the implications of item location for both equating and adaptive testing, Kingston and Dorans (1984) examined whether the difficulty of certain types of items is affected by exposure to the same item types earlier in the test. They found that for those item types in particular (analysis of explanations and logical diagrams) difficulty decreased substantially as familiarity with the item format increased. These effects, they concluded, were due, at least partially, to the complexity and novelty of the item types. For another type of item (reading comprehension), difficulty increased moderately as the item appeared later in the test. They attributed this outcome to fatigue.

Large location effects for even moderately speeded tests are ubiquitous. In a study of speededness, Wainer et al. (2004) found the unsurprising result that the more speeded the test, the greater the increase in difficulty for items situated toward the end of an SAT math test.

Cross-information

Another undesirable effect of context that may appear on a test is cross-information. *Cross-information* refers to information that one item may inadvertently contribute about the answer to another. For example, suppose the following item appears on a test for some individuals but not for others:

1. Carbon dioxide (CO_2) is a component of all of the following except
   a. seltzer b. ammonia c. "dry ice" d. photosynthesis
   but then some of those who answered this incorrectly, as well as some who never saw it, were presented with the easier question
2. The symbol for carbon dioxide is
   a. CO_2 b. H_2O c. NH_4 d. C_2O

Surely, those who had seen Question (1) would have an easier time with (2). Or, put in more general terms, the difficulty of (2) is dependent upon what preceded it. This is always true in test construction, but its effects are controlled for in two ways. First, test developers carefully construct tests to avoid such dependencies. And second, since everyone who takes a fixed test receives exactly the same questions, any such hints are fairly distributed – everyone gets the same one – and so no one is unfairly advantaged. With a test that is constructed by any algorithm that does not take into account the precise content of the items, dependencies among items can occur. This is a problem, but a relatively venial one. A test that is algorithmically constructed to be tailored to the individual examinee can yield dependencies that are unfairly distributed among examinees. This is a more serious problem.

Unbalanced content

A third type of context effect devolves from the content balancing of tests. All well-developed tests are built around content specifications. These are the content areas the test developer feels that the test ought to span. For example, in an arithmetic test, we might want to have 25% of the items deal with addition, 25% with subtraction, 25% with division, and 25% with multiplication. We will refer to specifications such as these as formal content specifications, since they deal with the formal contents of the subject area. There are also informal content specifications, which are usually not explicitly stated, but are both real and important. As an example of these, consider the structure of a problem in which the actual task is embedded. Suppose our arithmetic test consists of many "word problems" such as

3. If John caught 6 fish and threw 3 back and Mary caught 7 fish but threw 4 back, who brought home more fish?
   a. John   c. Both the same
   b. Mary   d. Can't tell from information given

Test developers have found that it is not wise to have too many problems dealing with the same topic (fishing); nor even the same general area. In the review of one test, one criticism was that there were too many "water items" (fishing, water skiing, boating, swimming, canoeing, etc.). While it may not be obvious why too many "water items" would be unfortunate, it is easily seen how some subgroups of the general population would be disadvantaged if there were many items on golf. Aside from the obvious social class effect, it also increases the potential for contributory information across items. Thus, in addition to filling specifications regarding the formal content, test developers must be careful to balance the test with respect to the informal contents. In a fixed-form test, it is straightforward (although not always easy) to be sure that the specifications are filled satisfactorily. Moreover, test developers can also read over the form carefully to assure themselves that there is no imbalance vis-à-vis the informal content.

It is not too hard to imagine how one could structure a computer algorithm to construct a test that would balance its content (e.g., van der Linden & Boekkooi-Timminga, 1989). The candidate items would be classified by their formal content and a categorical choice algorithm ("choose one from group A and two from group B") could be instituted. Of course, there may be need to cross-classify items by their difficulty as well, but this is just a bookkeeping task, and does not pose a complex technical problem, except for the need to write items in each content area that cover the entire range of difficulty. This does pose a problem on broad range tests, since, for example, it may not be easy to write sufficiently difficult arithmetic items or sufficiently easy calculus items. But this too can be surmounted if the unidimensionality assumption underlying IRT holds reasonably well: for in this case, the item's parameters are all that are required to characterize the item on the latent variable of interest.

It is more difficult to try to conceive of any categorization scheme that would allow a computer algorithm to determine if there was an overabundance of items on an inappropriate subject matter – where the subject matter was, in some sense, incidental to the item's content. To accomplish this requires either a finer level of item characterization (and hence a huge increase in the size of the item pool) than is now available or a level of intelligence on the part of the algorithm that is far beyond anything currently available.

Another type of content imbalance is in reference to gender, race, or ethnic groups across items. Such an imbalance violates current test development standards and, some claim, may serve to distract examinees and create feelings of hostility or resentment that could hinder their test-taking performance. An example of this might be a selection of a succession of items that use only female, or only male, references in professional roles.

4.2.2. Item difficulty ordering

A different kind of context effect, one that is not undesirable but rather is often planned and built into linear, nonadaptive tests, is the sequential ordering of items by their increasing difficulty. In the conventional testing situation, a test or section begins with items whose difficulties are less than all but the least able of the examinees; the difficulties increase as one progresses through the test, concluding with items whose difficulty exceeds the ability of all but the most able.

It is generally believed that beginning a test with the easiest items and ordering them by increasing difficulty allows examinees, especially those who are least able, to develop confidence at the outset of the test with the easiest items. They can then work through the test to their capacity in the time allowed, without wasting time or becoming unnecessarily discouraged by items that are far too difficult for them. This approach also tends to minimize guessing, because items on which those who are least able (and therefore most prone to guessing) spend the greatest amount of time are those on which they have the greatest likelihood of knowing the answer.

In an adaptive test, however, the initial item administered is usually one of moderate difficulty for the entire group being tested. Thus, for an examinee more able than average, the difficulty of successive items increases from this rather easy start, just as with a conventional linear test. For those at the low end of the ability continuum, however, the test begins with items that are excessively difficult. These become less difficult as the test progresses and incorrect responses accumulate, until the items emerge into a range of difficulty suitable for this group. But the items come at the examinees from above rather than below, contrary to traditional testing lore, suggesting that it may disadvantage lower ability groups. Indeed, there is considerable evidence, spanning half a century, that when items are administered from most to least difficult, rather than in the more usual order, the difficulty of the test as a whole increases (Hambleton, 1986; MacNicol, 1956; Mollenkopf, 1950; Monk & Stallings, 1970; Sax & Carr, 1962; Towle & Merrill, 1975).

4.2.3. Summary of context effect problems

In the nonadaptive testing situation, a test developer can read and study each test directly during its development phase to identify and mitigate context effects. Unfortunately, CAT makes such a strategy difficult or impossible. At present, the only way to ensure that cross-information is never inadvertently provided on a CAT is to compare all conceivable subsets of items derivable from the item pool. This is at best an unreasonable undertaking, especially because a viable item pool must contain at least hundreds of items and is continuously being updated, with new items being added and old ones retired. The identification and elimination of the other context effects just discussed is even more difficult under the traditional adaptive framework. It would require an analysis of the content of each and every test produced by the CAT item-selection algorithm. This is impractical with current technology.

4.2.4. An issue of robustness

The increased efficiency that is the hallmark of adaptive testing – the same accuracy of measurement as a linear test at a greatly reduced length – is not an unalloyed good. The shorter test lacks the redundancy inherent in the almost twice as long conventional linear test and so becomes more vulnerable to idiosyncrasies in item performance. As a result, when an item is failing to perform in accordance with its parameter estimates on a CAT, the detrimental impact on validity is about twice what would result from a traditional nonadaptive test of about the same theoretical accuracy – but of greater length (Green, 1988). Furthermore, if the flawed item is one of moderate difficulty and relatively high discrimination, its detrimental influence is likely to be felt many times because most CAT item-selection algorithms favor such items (Wainer, 2000).

4.3. Traditional solutions

Unfortunately, there is no obvious solution for all of the variety of problems related to implementing CAT that we have just discussed. However, certain avenues of investigation invite exploration in the search for solutions. Most of these avenues involve a more variegated stratification of the item pool. Thus, the statistical item-selection algorithm would be limited to choosing the most informative item from within rigidly controlled strata. Inexorably, this means generating an item pool of substantially greater size than was initially envisioned by CAT enthusiasts.

4.3.1. Context effects

It may be possible to stratify finely enough to avoid cross-information, to allow the balancing of gender, race, and ethnic references, and to prevent redundancy of item topics. Such stratification does nothing to improve the fungibility of the items. Items still remain context dependent. As long as different examinees are being compared to one another on the same items in different contexts, this problem will remain. It does not seem likely that the solution to this is statistical; trying to adjust an item's parameter values on the basis of its (possibly) unique circumstance is beyond current technology.

4.3.2. Robustness

The increased influence of a single item within a CAT framework is a double-edged sword. while a flawed or inappropriate item can have an undue influence on the ability estimate, an additional corrective item can get us back on the right path doubly quickly. Thus, we feel that it is important to emphasize that CAT users ought to leave in some redundancy so as to allow for an unforeseen flaw. It is probably wise to choose a conservative estimator for the standard error of estimates of ability as well, since many CATs have stopping rules based on this criterion. This avoids stopping an examination "too soon" due to possibly incorrect items that could have the unhappy consequence of an ability estimate that is not as precise as required.

4.3.3. Item difficulty ordering

The effect of the order in which items are presented can be ameliorated by choosing a starting point for the CAT that is toward the low end of the ability continuum. The lower this is, the less the order effect problem, for more and more examinees will have the items ordered from easy to difficult. However, the lower the starting value, the less efficient the CAT algorithm. This may be a relatively venial problem if big jumps are made initially in item choice, which is consistent with most CAT item-selection algorithms.

4.4. Testlets – an alternative solution

The problems that arise from adaptive testing can be eased through an alternative approach. The key idea is to use a multi-item testlet as the fundamental unit of test construction and test administration. We shall define a *testlet* as a group of items that may be developed as a single unit that is meant to be administered together. Although the path through a testlet could be branched, our focus, at least for now, is on linear testlets that contain n items, where n could be as few as one, but more typically would have four or more items. All examinees who are presented with a particular testlet would be confronted with the same items in the same order.

One example of a testlet is the traditional reading comprehension item type in which the examinee is presented with a passage and a bundle of items related to that passage. In similar vein, science testing commonly uses a graph or table as the focus of a set of related items, and many history and geography tests use a map as the central stimulus for a set of items.

How does this concept help us to resolve the problems stated previously? First, although a testlet-based test can be made adaptive by hierarchically structuring the testlets, the path through the testlet itself can be carefully examined to ensure that it does not suffer from the same problems as items within a standard CAT. For example, the items within a testlet can be ordered from easy to difficult. Thus because order effects are localized, their effects are greatly diminished. This is roughly akin to the way that on linearly administered tests, test sections are difficulty ordered, but when the examinee moves to a new section, the items within it are ordered from easy to difficult. Obviously, most testlets are shorter than a typical test section, but the general idea is the same.

In addition, when blunders do occur with the testlet model, they can be more easily overcome, since all examinees who receive a particular testlet receive exactly the same testlet. Thus any blunder that occurs, occurs repeatedly and can be more easily detected, and test scores of those affected adjusted accordingly.

Of course, testlets have other advantages, quite apart from those associated with providing greater control within an adaptive framework. By far, their greatest value is that they provide another formal mechanism for test developers to test the construct in a way that makes the most sense.

In the late 1980s, when adaptive testing was still in its infancy, the first large-scale adaptive test was the adaptive version of the Armed Services Vocational Aptitude Battery (CAT-ASVAB). The ASVAB, fully described by Sands, Waters, and McBride (1997), is used by the military to classify and select about two million young men and women every year, making it the largest single testing program in the world. By making the ASVAB adaptive, substantial savings in time and money were realized. But the test had to have a few modifications in its transition to computer-based administration. One key change, important for this discussion, was in the section on paragraph comprehension. The CAT-ASVAB was scored using traditional IRT. As we mentioned in

Chapter 1 and discussed in greater detail in Chapter 3, a key requirement for IRT to hold is that the items of the test should be conditionally independent (otherwise Equation (3.4) would not be valid). Yet, it was clear that traditional paragraph comprehension items, because they are connected together with a common passage, are not likely to be conditionally independent. This put the developers of the CAT-ASVAB into a quandary. If they left the paragraph comprehension section alone, the scoring model was very likely to be incorrect. But how to change it? The typical passage was about 300 words long and was accompanied by five items.[1] If they just used a single item, the local dependency problem would go away. But getting only one item's worth of information after asking an examinee to spend so much time reading and figuring out a long passage seemed wasteful of effort. It was also antithetical to the efficiency goals that are fundamental to adaptive testing. What to do? It was decided to ask but a single question while shortening the typical paragraph to about 120 words.[2] The new items showed an increase in their correlation with word knowledge and correlated only 0.38 with the prior form of paragraph comprehension items administered in the paper-and-pencil format (ASVAB P&P form 10B – Moreno & Segall, 1997, p. 172). The median correlation of all of the other ASVAB subtests with their CAT counterparts was 0.72. Thus, the shortening of the passage solved the immediate problem, but, as it turned out, changed the construct being measured.

If, instead, the passage and its associated items were thought of not as a set of independent items but rather as a testlet, to be administered as a unified whole, the problem would have solved itself. But how?

In 1988, Paul Rosenbaum made an important distinction. He pointed out that local independence could be violated because of idiosyncratic features of the test format (e.g., the same passage associated with several items) or because of actual departures from the unidimensionality of the test construct. Rosenbaum (1988) proved (Theorem 1) that given the loss of conditional independence within testlets, unidimensionality can still prevail between testlets. Hence, if we use the testlet as the unit of measurement, a unidimensional item-response model may still be appropriate.

Thus, Rosenbaum's theorem permits the use of items with desirable characteristics, with less concern for violation of the assumption of conditional independence. Hence, traditional paragraph comprehension items, when scored as testlets, can be used without additional concern by the test developer.

[1] P&P ASVAB forms 23, 24, 25, and 26 had 5 items per passage, and the typical passage ranged up to 326 words (John Welsh, personal communication, July 9, 2004).

[2] CAT ASVAB forms 1, 2, 3, and 4 had 1 item per passage, and the typical passage had no more than 120 words (John Welsh, personal communication, July 9, 2004).

One of the strengths of Rosenbaum's result is that if a traditional IRT model shows a lack of fit when applied to testlet data, a statistical test using contingency table methods can be applied to determine whether the lack of fit is due to a loss of conditional independence within testlets. If this is found to be the case, examinee performance can still be summarized with a single parameter, as the between-testlet scoring may still satisfy the requirements of a unidimensional model.

4.5. An initial summing up

The problems discussed in this chapter are not certain to occur within the context of adaptive testing, nor are they guaranteed to be cured through the use of testlets. Rather, we believe that the bundling of items into testlets provides a mechanism through which the likelihood of these unfortunate circumstances will be reduced without seriously diminishing the efficiency of an adaptive test and will allow the test's structure to more closely match the constructs that the users of the test want to draw inferences about. More specifically:

1. Context effects
   a. *Cross-information:* By enlarging the fungible unit of the test from the item to the testlet, we are reducing the "boundary effects" of presentation. Only the first item in the testlet has an unknown predecessor; all the others follow an item that experts have judged to satisfy the canons of good practice. These boundary effects include such things as cross-information, which, though not entirely eliminated, has had its likelihood reduced substantially. Moreover, if it does occur, the contaminating item is likely to be more distant and so reduces its effect.
   b. *Unbalanced content:* By controlling the presentation more tightly, we can better ensure that the test specifications are satisfied for each individual test. For example, we know of an adaptive science test that consisted of biology, chemistry, and physics questions. The physics questions turned out to be somewhat more difficult than either of the other areas. One student, who studied biology and physics, but not chemistry, found that the questions presented to him vacillated between biology (which he tended to answer correctly) and chemistry (which he tended to miss). He never got to demonstrate his expertise in physics. Such a situation reflects the lack of unidimensionality within the test and could be solved within a traditional CAT framework, but its solution is accomplished so much more gracefully with three testlets in series.

2. *Robustness:* Once again, this is a double-edged sword. If a flawed item appears in a testlet, measurement accuracy will be affected, but fairness can still be served. With a testlet, examinees whose abilities are near one another will receive similar tests. The closer together they are, the closer together their tests will be.[3] Thus, the appearance of a flawed item will tend to have a diminished impact on individual comparisons. CATs provide protection by allowing recovery from what turns out to be an anomalous response. Testlets also provide protection, because they are constructed as a whole and the plausible interplay among its members could be subjected to expert judgment in preparation, and statistical quality control in pilot testing. Also, because a testlet is self-contained, any test construction errors that do show are localized and the offending part can be replaced without overly disturbing the rest of the test.
3. *Order effects:* Item difficulty ordering can be controlled by controlling starting points. Even if this is not done, we once again obtain fairness because of the testlet characteristic of having those whose ability estimates are close having closely matching tests.

The strength of the testlet approach derives from two of its characteristics. The first is that it allows greater control of item presentation. The benefits of this were just elaborated upon. The second is that it enforces the notion that the more difficult it is to discriminate between two examinees, the more identical the test they took is likely to have been. Thus, while we may still have problems with multidimensionality – flawed items and other deviations from the test model – those problems will be the same for the particular individuals being compared. We do not mean to imply that test equating does not yield valid scores that allow comparisons across different test forms. This is certainly true to some extent. But the legitimacy of such comparisons depends on the extent to which all of the assumptions of the test scoring model and the equating method are upheld in the data; the smaller the difference between the two individuals being compared, the greater the likelihood that the standard error of the scores will affect the validity of the comparison. When two individuals take precisely the same test, the error of equating is removed from the mix (and substituted for it are the ills associated with problems of security – if one person took the test

[3] With the security precautions now common, it is possible that two examinees with identical abilities can, in theory, get two very different but parallel test forms. But this outcome relies on an item pool of substantial size – indeed, it needs to be very large at all levels of ability. Even for tests with hundreds of thousands of examinees, and with the resources that such volume provides, test items are sparse at some levels of ability, and very few items account for most of the administered tests (Wainer, 2000). Thus, for tests constructed within more modest circumstances, it is likely that individuals with similar abilities will receive a substantial number of identical items.

well before the other, there is always the chance of some sort of security leak that would allow the later examinee to have access to the items).

Testlets will not cure this, but they can help to ameliorate it by ensuring that if two examinees' test forms share a common testlet, they will share all items in the testlet. This means that, for those items, no equating is required, and hence the error of equating is reduced.

The use of testlets also requires a shift in our thinking about psychometric modeling. True score theory (Lord & Novick, 1968) might also be called "test response theory" since it uses as its fungible unit something that is too large for adaptive testing – the whole test. Traditional summary statistics, such as true score, test reliability, and test validity focus on the entire test as the testing unit. Item response theory focuses on the item and it produces statistics, such as item difficulty, which are points on an underlying continuum. The examinee's score on the test is also estimated as a point on the same continuum. These estimates and their associated measures of precision stand alone and are not referred to the specific items taken; the underlying trait is all that matters. We are recommending a middle path that we term "testlet response theory" since it uses, as its unit of measurement, pieces of the test that are simultaneously small enough to be usefully adaptive and large enough to maintain some stability. Tests have long been divided into different sections; our contribution is to suggest a more flexible unit, the testlet. The testlet can be as small as a single item (although in this extreme case, none of the advantages discussed here would hold), as large as the entire test, or anything in-between. Our point is that the size of the testlet should be determined by the character of the constructs being measured, not the available psychometric theory. In the balance of this monograph, we describe formal psychometric theories that can accommodate testlets of any character.

Questions

1. What testing problems was the invention of the testlet meant to address?
2. Why was the solution to these problems made more urgent with the popularization of CAT?
3. How does the use of testlets ameliorate these problems?
4. What crucial role did Rosenbaum's theorem on the independence of item bundles play in TRT?

References

Eignor, D. R., & Cook, L. L. (1983, April). *An investigation of the feasibility of using item response theory in the preequating of aptitude tests.* Paper presented at the annual meeting of the American Educational Research Association, Montreal, Canada.

Green, B. F. (1988). Construct validity of computer-based tests. In H. Wainer & H. Braun (Eds.), *Test validity* (pp. 77–86). Hillsdale, NJ: Erlbaum.

Hambleton, R. K. (1986, February). *Effects of item order and anxiety on test performance and stress*. Paper presented at the annual meeting of Division D, the American Educational Research Association, Chicago.

Kingston, N. M., & Dorans, N. J. (1984). Item location effects and their implications for IRT and adaptive testing. *Applied Psychological Measurement, 8*, 146–154.

van der Linden, W. J., & Boekkooi-Timminga, E. (1989). A maximin model for test design with practical constraints. *Psychometrika, 54*, 237–248.

Lord, F. M. (1971a). The self-scoring flexilevel test. *Journal of Educational Measurement, 8*, 147–151.

Lord, F. M. (1971b). Robbins-Munro procedures for tailored testing. *Educational and Psychological Measurement, 31*, 3–31.

Lord, F. M. (1974). Estimation of latent ability and item parameter when there are Omitted responses. *Psychometrika, 39*, 247–264.

Lord, F. M. (1980). *Applications of item response theory to practical testing problems.* Hillsdale, NJ: Erlbaum.

Lord, F. M., & Novick, M. R. (1968). *Statistical theories of mental test scores*. Reading, MA: Addison-Wesley.

MacNicol, K. (1956). *Effects of varying order of item difficulty in an unspeeded verbal test*. Unpublished manuscript. Educational Testing Service, Princeton, NJ.

Mollenkopf, W. G. (1950). An experimental study of the effects on item analysis data of changing item placement and test-time limit. *Psychometrika, 15*, 291–315.

Monk, J. J., & Stallings, W. M. (1970). Effect of item order on test scores. *Journal of Educational Research, 63*, 463–465.

Moreno, K. E., & Segall, D. O. (1997). Reliability and construct validity of CAT-ASVAB. In W. A. Sands, B. K.Waters, & J. R. McBride (Eds.), *Computerized adaptive testing: From inquiry to operation* (chap. 17, pp. 169–174).Washington, DC: American Psychological Association.

Rosenbaum, P. R. (1988). A note on item bundles. *Psychometrika, 53*, 349 –360.

Sands, W. A., Waters, B. K., & McBride, J. R. (Eds.). (1997). *Computerized adaptive testing: From inquiry to operation.* Washington, DC: American Psychological Association.

Sax, G., & Carr, A. (1962). An investigation of response sets on altered parallel forms. *Educational and Psychological Measurement, 22*, 371–376.

Towle, N. J., & Merrill, P. F. (1975). Effects of anxiety type and item difficulty sequencing on mathematics test performance. *Journal of Educational Measurement, 12*, 241–249.

Wainer, H. (2000). Rescuing computerized testing by breaking Zipf's Law. *Journal of Educational and Behavioral Statistics, 25*, 203–224.

Wainer, H., Bridgeman, B., Najarian, M., & Trapani, C. (2004). How much does extra time on the SAT help? *Chance, 17*(2),17–23.

Wainer, H., Dorans, D. J., Eignor, D., Flaugher, R., Green, B. F., Mislevy, R. J., Steinberg, L., & Thissen, D. (2000). *Computerized adaptive testing: A primer (2nd ed.).* Hillsdale, NJ: Erlbaum.

Wainer, H., Dorans, D. J., Flaugher, R., Green, B. F., Mislevy, R. J., Steinberg, L., & Thissen, D. (1990). *Computerized adaptive testing: A primer.* Hillsdale, NJ: Erlbaum.

Wainer, H., & Kiely, G. L. (1987). Item clusters and computerized adaptive testing: A case for testlets. *Journal of Educational Measurement, 24*, 185–201.

Weiss, D. J. (1974). *Strategies of adaptive ability measurement* (Research Report 74-5). Minneapolis: University of Minnesota, Psychometric Methods Program.

Weiss, D. J. (1982). Improving measurement quality and efficiency with adaptive testing. *Applied Psychological Measurement, 6*, 473–492.

Weiss, D. J., & Kingsbury, G. G. (1984). Application of computerized adaptive testing to educational problems. *Journal of Educational Measurement, 21*, 361–375.

Whitely, S. E., & Dawis, R. V. (1976). The influence of test context effects on item difficulty. *Educational and Psychological Measurement, 36*, 329–337.

Yen, W. (1980). The extent, causes, and importance of context effects on item parameters for two latent trait models. *Journal of Educational Measurement, 17*, 297–311.

5

The origins of testlet response theory – three alternatives

As we mentioned earlier, the concept of the testlet was explicitly introduced by Wainer and Kiely (1987, p. 190) as "a group of items related to a single content area that is developed as a unit and contains a fixed number of predetermined paths that an examinee may follow." In Chapter 4, we proposed the use of the testlet as the unit of construction and analysis for computerized adaptive tests (CATs) with the expectation that they could ease some of the observed and prospective difficulties associated with most current algorithmic methods of test construction. Principal among these difficulties are problems with context effects, item ordering, and content balancing.

We can summarize the way that testlets can help into two broad categories: control and fairness.

Control, in the sense that by redefining the fungible unit of test construction as something larger than the item, the test developer can recover some of the control over the structure of the finished test that was relinquished when it was decided to use an automatic test-construction algorithm.

Fairness, in the sense that all examinees who are administered a particular testlet, in addition to getting a sequence of items whose content and order have been prescreened and approved by a test developer and an associated test-development process, also get the same sequence as other examinees whose observed proficiencies are near theirs. Thus, when comparisons among examinees of very similar proficiency are made, those comparisons will be made on scores derived from tests of very similar content.

Since the introduction of testlets, a number of prospective uses have emerged. Some of these were foreseen in the original presentation, some were not. In this chapter, we describe three uses of testlets, each of which illustrates a different kind of testlet construction. We then derive psychometric models that can be used to score tests composed of testlets in each of these forms. We do not, by our categorization here, mean to imply that these are the only ways that testlets can be constructed, nor are these the only psychometric models that will prove to be efficacious. Rather, we present these as reasonable ways to begin to think about a psychometrics for testlets.

The testlet scoring models we present are quite different psychometric characterizations from one another, and each could presumably be helpful in some special circumstance. Therefore, rather than being prescriptive at this point, we have chosen to try to be expansive. Actually, it is our current belief that the psychometric characterization that begins in Chapter 7 is likely to serve the broadest range of needs, but, depending upon the actual situation, sometimes different methods will be needed in different circumstances, to serve one's needs best.

Example 1.. *NCARB – Content balanced testlets, randomly selected – scored with standard IRT combined with a sequential probability ratio test*
(Lewis & Sheehan, 1990)

The National Council of Architectural Registration Boards (NCARB) commissioned the Educational Testing Service (ETS) to develop a "test of seismic knowledge" for use in the architectural certification process. There was a broad range of knowledge and experience among the prospective examinees. Some were so expert in these matters that even a cursory oral examination would reveal this quite clearly. On the other extreme, there were some examinees whose knowledge of this area was so sparse that this too would be revealed with only a quick look. However, many prospective examinees fell comfortably between these two extremes, and so required an examination in more depth. It was hoped that a test could be constructed that would adapt itself to the level of discrimination required. If the judgment was an easy one ("pass" or "fail"), it would discriminate with alacrity (using as few items as required for the precision of measurement needed). If the judgment was more difficult, the test would lengthen itself to yield the required accuracy. Thus, it was desired to have a test that as questions were asked of the examinee, and their responses noted, the testing algorithm would continually ask itself, "Pass?" "Fail?" "Keep testing?" This determination would be made on the basis of a Bayesian decision process in which a loss function was constructed on the basis of the costs associated with (1) passing someone who should have failed, (2) failing someone who should have passed, or (3) asking more questions.

To be able to utilize the powerful statistical machinery of sequential decision making in building a test with an adaptive stopping rule, it was critical that all of the test's component parts be exchangeable. That is, that any potential piece of the test in the pool can be substituted for any other piece, with no degradation of the test. It was also important that the test that any examinee is administered be comparable, in terms of difficulty and content, to the test that any other examinee received, regardless of their respective lengths. Obviously, trying to develop an item pool in which all of the items are exchangeable is at least a difficult task, and probably an impossible one. Testlets proved to be a useful tool in the successful completion of this ambitious application since, as we shall soon show, they can be constructed to be exchangeable.

There was an existing item pool of 110 items that had previously been administered to another sample of the examinee population in a paper-and-pencil format. Their responses to these items had previously been fit with the three-parameter logistic model (3-PL) and these estimated parameters were available (Kingston, 1987). The characteristic functions of these 110 items varied in difficulty and slope. They also fell into one of two content categories. Sixty percent of the items were Type I items that dealt with physical and technical aspects of seismic knowledge. Forty percent were Type II items that dealt with economic, legal, and perceptual aspects of seismic knowledge.

It was decided to divide the item pool into 10-item testlets in which each testlet was balanced for content and equal in average difficulty and discrimination. The difficulty of each testlet was aimed at the cut score for the test, and so a simple item-presentation algorithm would then suffice to achieve the desired ends. Specifically, if such interchangeable testlets could be constructed, one could choose any testlet at random for presentation to an examinee. After the completion of that testlet, a statistical determination could be made to pass the examinee, fail him, or present another testlet. Since all testlets would be content balanced, we could be sure that all examinees, regardless of the length of the test that they received, would have received a test of identically balanced content. Moreover, since all testlets were of the same average difficulty, we could be sure that no examinees were unfairly advantaged (or disadvantaged) with a too easy (or too difficult) exam. Last, since all testlets were of equal average discrimination, the precision of all tests was strictly proportional to their length (and not of their particular makeup). This latter aspect is of importance for some technical issues dealing with the calculation of the posterior expected loss. When testlets are constructed in this way, the number-correct score carries all of the information we need to be able to implement the Bayesian decision process that was employed in this application.

The testlets were constructed by cross-classifying the item pool according to (a) type designation, and (b) estimated item difficulty. Next the items were assigned sequentially to testlets. Each resulting testlet had six Type I items and four Type II items. The six testlets that appeared most "parallel" (in terms of average difficulty and discrimination) were then selected from the 11 available. The obtained testlets were then examined by test developers who were experts in seismology. The validity of the testlet interchangeability assumption was evaluated by determining the degree to which the six selected testlets varied with respect to the average likelihood of a particular number-correct score. Likelihoods were evaluated at five different points on the latent proficiency scale. These five points corresponded to important decision points surrounding the anticipated cut score. This validity check showed that for examinees near the cut score, the average number-correct score had the same probability regardless of which testlet was administered. This strongly supported the assumption of "parallel testlets" for examinees whose number-correct score was in the regions examined. That such parallel testlets, with balanced content, could be so easily constructed provides us with a powerful and practical tool for building fair and balanced tests that have an adaptive stopping rule.

The NCARB test is being employed currently with a minimum test length of 20 items (two testlets) and a maximum of 60 items (six testlets). The testlet building blocks make it possible to utilize easily the powerful machinery of sequential probability tests that was originally developed by Abraham Wald (1947) for quality control problems. It allows us to maximize the probability of correctly classifying individuals while minimizing the amount of testing that must be done.

Example 2.. *Reading Comprehension – linear testlets, linearly administered – scored with a polytomous IRT model*

(Thissen, Steinberg, & Mooney, 1989)

Reading comprehension items, formed by a single passage followed by a number of related questions, have long been a common and useful component of most verbal tests, yet, as discussed in Chapter 4, their use has been restricted within IRT-scored CATs. The reason for this is that it was (properly) recognized that the (sometimes) crucial assumption of conditional independence among items was likely to be violated if several items share the same stem. There have been three different responses to this problem. One, decided upon by the development team for the Computerized Adaptive Test version of the Armed Services Vocational Aptitude Battery (CAT-ASVAB) was to modify the items so that each passage is accompanied by a single item only. As discussed previously, this solved the issue of loss of conditional independence, but at a considerable cost in efficiency – one gets only a single response's worth of information from the time taken to read an entire passage. The response to make the passages shorter increased the efficiency, but changed the trait that was being measured by the test. A second approach, quite commonly taken, is to ignore the interdependencies among the items associated with a single passage, fit a binary response model, and hope that everything is okay. This approach tends to overestimate the information yielded by the item. The third approach, which is moving in a useful direction, is a testlet approach that we shall describe in this section. In it, we explicitly define the passage with its m associated questions as a single item; or, more precisely, a single testlet. In this formulation, we will consider the examinees' responses to the m questions as a polytomous response, and then score it either 0, 1, 2, ..., or m, depending upon how many of the m questions the examinee gets correct.

Using this approach, we are able to reduce concerns about the lack of conditional independence among the questions associated with a single passage. The model we use (developed by Bock, 1972) does not require conditional independence within testlets, but only between them. This latter requirement involves independence between passages (after conditioning on examinee proficiency). If the dependencies are only within testlet, and not endemic throughout, Rosenbaum's (1988) result assures us that polytomous response models, one of which we shall describe shortly, can represent the data properly. This approach was a substantial improvement over what was used previously, but it too will be superceded by the models we develop in Part II.

5.1. Bock's model

Suppose we have J testlets, indexed by j, where $j = 1, 2, \ldots, J$. On each testlet, there are m_j questions, so that for the jth testlet there is the possibility for the polytomous response $x_j = 0, 1, 2, \ldots, m_j$. The statistical testlet scoring model posits a single underlying (and unobserved) dimension, which we shall continue to call *latent proficiency*, and denote as before, θ. The model then represents the probability of obtaining any particular score as a function of proficiency. For each testlet, there is a set of functions, one for each response category. These functions are sometimes called item characteristic curves (Lord & Novick, 1968), item-operating curves (Samejima, 1969), or trace lines (Thissen, Steinberg, & Mooney, 1989). We shall follow Thissen et al.'s notation and nomenclature.

The trace line $P(T_{jx} = m_j)$ for score $x = 0, 1, \ldots, m_j$, for testlet j is

$$P(T_{jx} = m_j) = \frac{\exp[a_{jk}\,\theta + c_{jk}]}{\sum_{k=0}^{m_j} \exp[a_{jk}\,\theta + c_{jk}]}, \tag{5.1}$$

where $\{a_k, c_k\}_j$, $k = 0, 1, \ldots, m_j$, are the item category parameters that characterize the shape of the individual response trace lines. The model is not fully identified, and so we need to impose some additional constraints. It is convenient to insist that the sum of each of the parameters equal 0, that is,

$$\sum_{k=0}^{m_j} a_{jk} = \sum_{k=0}^{m_j} c_{jk} = 0. \tag{5.2}$$

This model was fit to a four-passage, 22-item test of reading comprehension by Thissen et al., in which there were seven items associated with the first passage, four with the second, three with the third, and eight with the fourth (or, in the notation just introduced, where m_j takes on the values 7, 4, 3, and 8). They did this after they had performed an item-factor analysis (Bock, Gibbons, & Muraki, 1988) and found that a multifactor structure existed. The (at least) four-factor structure found among these 22 items made the unidimensional (conditional independence) assumption of traditional IRT models untenable. After considering the test as four testlets and fitting Bock's nominal response model to the data generated by the almost 4,000 examinees, they compared the results obtained with what would have been the case if they had ignored the lack of conditional independence and merely fit a standard IRT model. They found two things. First, that there seemed to be a slightly greater validity of the testlet derived scores. Second, the test information function yielded by the traditional analysis was much too high. This was caused by this model's inability to deal with the excess intrapassage correlations among the items (excess after conditioning on θ). The testlet approach thus provided a more accurate estimate of the accuracy of the assessment.

Example 3.. *An algebra test – using hierarchical testlets, linearly administered – uses validity-based scoring*

Mathematical knowledge has a partially hierarchical structure that lends itself to the construction of hierarchical testlets. Two testlets related to elementary algebra were prepared by ETS Test Development staff members James Braswell, Jeanne Elbich, and Jane Kupin. In Figure 5.1 is a sample testlet that might form part of a larger test consisting of a number of such testlets, each of which covers a different topic in a broader mathematical unit. These testlets would be administered linearly, that is, each student would respond to items in all testlets, so that balancing of content occurs between testlets. A given testlet could focus on the content related to what are felt to be important topics in the subject.

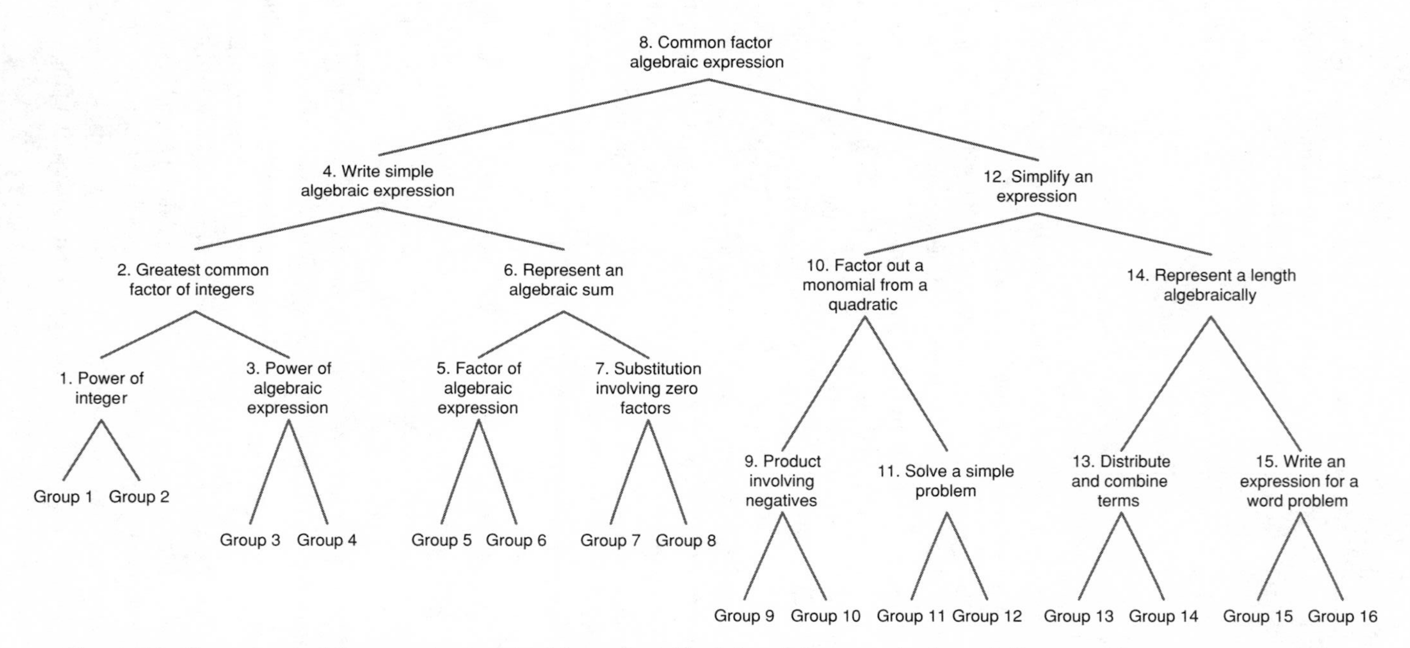

Figure 5.1. Elementry algebric expression testlet – 15 questions. The left path from a node always indicates an incorrect responses to the question represented by the topic shown at that node.

The idea of the hierarchical testlet is that the student is routed through the items according to his or her performance. After a correct answer, an item addressing a more difficult concept is presented. After an incorrect answer, an item testing a less difficult concept is given.

As an illustration, consider a student who answers the first item in Figure 5.1 (identifying the greatest common factor in two algebraic expressions) incorrectly. This student would then be asked to perform the theoretically simpler task of writing a simple algebraic expression. Suppose the student is able to answer this item correctly. In this case, the student would next be asked to represent an algebraic sum. A correct answer here and the fourth item presented to the student would require a substitution involving zero factors. The actual items that this hypothetical student would receive are presented in Figure 5.2.

At the end of the sequence, students have been grouped into 16 theoretically ordered levels, on the basis of the patterns of their responses. Conceptually, such a test is appealing, since, under certain conditions, it provides the same resolution among the students taking the test as does the "number correct score" of a 15-item test, with each student taking only four items. It does, however, present some difficult questions about how to model responses and estimate proficiency. In addition, while the resolution of the test is the same as that of a 15-item test, its precision is not. We need to characterize this precision in a way that is quite different from that done by the traditional methods (IRT or classical reliability).

An alternative to dealing with these questions is offered by an approach known as *validity-based scoring* (Lewis, still in preparation). This approach is based on the assumption that it is possible to obtain information on some criterion measure (or measures), at least for a calibration sample of students. In the present context, this information might consist of scores on a battery of longer tests of various algebraic skills, or perhaps of teacher ratings on such skills that were based on a broad sample of student behaviors.

The grouping of students based on their responses to the items in the testlet may be expressed using indicator (0, 1) variables, one for each group. These indicator variables are then used as predictors for the criterion measures in the calibration sample of students. *Validity-based scoring* assigns the predicted values on the criterion as the scores for each possible outcome for the testlet. These scores are simply the mean criterion values for the group of students with each given testlet result. The group standard deviations on the criterion variables may be interpreted as conditional standard errors of measurement for these scores. If the criteria are indicator variables, as would be the case if teacher ratings assigned each student to one of several skill levels, then the scores for a given testlet outcome are the estimated conditional probabilities of being in each skill level, given testlet performance.

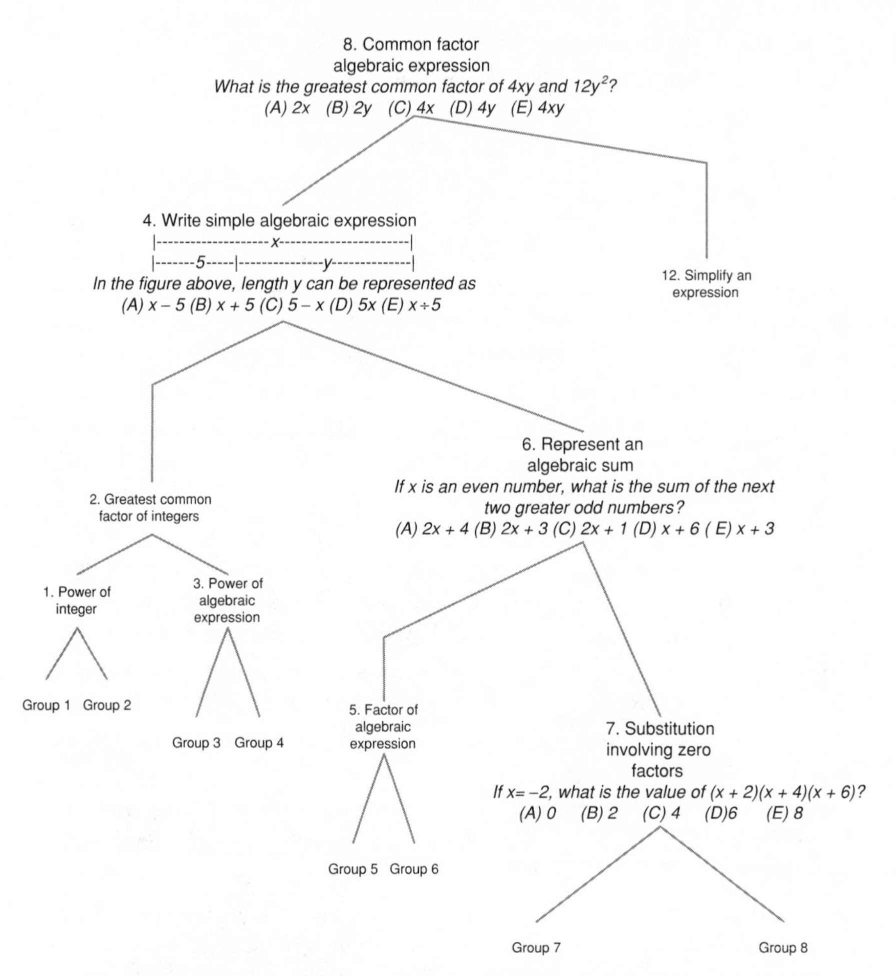

Figure 5.2. A sample sequence of items (and responses) from the testlet represented in Figure 5.1. The left path from a node always indicates an incorrect response to the question represented by the topic shown at that node.

Figure 5.3 illustrates this idea for the algebra testlet of Figure 5.1. Imagine a calibration sample of students, all of whom had responded to the algebra testlet, and for all of whom a criterion score (denoted by C) was available. The group of students (Group 1 in Figure 5.3) who incorrectly answered all questions presented to them (Items 1, 2, 4, and 8) have a mean score on the criterion, which is denoted by $\overline{C}_1$. This is the score that will subsequently be given to any student with this response pattern. Similarly, mean criterion scores for each of the remaining 15 groups in the calibration sample ($\overline{C}_2$ through $\overline{C}_{16}$) will be used as scores for students with any of the remaining response patterns. The

Group	Pattern	Item															Criterion
		1	2	3	4	5	6	7	8	9	10	11	12	13	14	15	score
1	0000	0	0		0				0								C_1
2	0001	1	0		0				0								C_2
3	0010		1	0	0				0								C_3
4	0011		1	1	0				0								C_4
5	0100				1	0	0		0								C_5
6	0101				1	1	0		0								C_6
7	0110				1		1	0	0								C_7
8	0111				1		1	1	0								C_8
9	1000								1	0	0		0				C_9
10	1001								1	1	0		0				C_{10}
11	1010								1		1	0	0				C_{11}
12	1011								1		1	1	0				C_{12}
13	1100								1				1	0	0		C_{13}
14	1101								1				1	1	0		C_{14}
15	1110								1				1		1	0	C_{15}
16	1111								1				1		1	1	C_{16}

Figure 5.3. Responses patterns, the item responses yielding them, and their associated criterian scores.

standard deviation of the observed C_i, within each group, serves as the standard error of the score.

Validity-based scoring provides a direct check on the theoretical ordering of the response groups. If the scores for the groups do not reflect their ordering, or if differences between scores for adjacent groups are small relative to the standard errors, follow-up diagnostics should be explored. One possible diagnostic procedure is to apply *validity-based scoring* at each hierarchical level of the testlet and see at what point the expected ordering begins to break down. This would then provide aid in the redefining of the hierarchical structure (and the associated items) that characterize both the theoretical structure of the testlet and its operational realization.[1]

5.2. Summary and conclusions

Testlets, as the name implies, are small tests that were first proposed as convenient units from which to construct a test. They are usually small enough to manipulate but big enough to carry their own context. They can be used to

[1] Among some other advantages of validity-based scoring are (i) a test can't be scored unless a clear criterion has been determined, (ii) the same test can yield different scores for different criteria, and (iii) tests cannot be used for purposes that do not suit it. In this latter case, because there is no relationship between performance on the test and the (improper) criterion, all examinees will be assigned the same score regardless of their performance on the test. Thus, if we used a test of height to determine examinees' mathematical ability, we would find a zero correlation and so the predicted mathematical score would be the same regardless of the candidates height. Which is as it should be.

guarantee content balance to an algorithmically constructed test by being sure that each testlet is balanced (we refer to this as "within testlet balancing" as in the NCARB example), or by letting each testlet span one aspect of the test specifications, so that the test contents are balanced by judicious choice of testlets. We refer to this as "between testlet balancing" (as in the algebra test example).

In this chapter, we have described three different ways that testlets can be used to improve the quality of tests and the flexibility of their mode of construction. While the original formulation of the testlet was phrased within the context of adaptive testing, we have shown that they can be useful in other situations as well. A testlet formulation provides us with a more accurate estimate of test quality in the paragraph comprehension example; it allows us to use the powerful statistical machinery of sequential decision making in the NCARB example, and it provides us with a much more efficient test in the algebra example. Moreover, they do this without the risks associated with putting the entire task of test construction into the hands of an imperfect algorithm.

The three examples we elaborated here were chosen for several reasons. One of these is that they illustrate the role that IRT can play in testlet construction and use. Specifically, IRT and testlets are two notions that are somewhat independent. One can use the testlet approach, even in an adaptive mode, without recourse to IRT at all (the algebra test). Or, one can tie the testlet's construction and scoring intimately to IRT (the paragraph comprehension test). Or, it can be in-between, where IRT is used to construct the testlets, but is then not used in the scoring (the NCARB example).

Obviously, there are many IRT models that can appropriately be utilized for the representation of testlet difficulty and for testlet scoring. The paragraph comprehension example only touches the surface of such models. More information could have been gleaned from the NCARB test if the full response pattern had been used rather than merely the number right within testlet. But how can this be accomplished while still accounting for the lack of local independence within testlet? This is the topic of much of the rest of this book. In the example provided here, such a refined scoring model was unnecessary, because the currently configured NCARB test is as efficient and accurate as required.[2]

In addition, there are other modes of testlet construction and combination that we have not touched upon. One of the most obvious ones would be testlets (either hierarchically or linearly formed) combined hierarchically. We leave a description of the scoring of such tests to a later account.

[2] Indeed, when this test was first developed, it was shown that for some individuals one testlet was sometimes sufficient to reach a decision with the desired accuracy; however, the members of NCARB felt that the minimum test length should be at least two testlets (20 items) so that examinees would believe that they were getting fair value for their testing fee.

We would like to emphasize one aspect of the algebra test's scoring scheme: the way that the scoring of the testlet is integrated fully with a validity criterion. We feel that this is very important, for the validity of a test's scores is their most important characteristic and that too often validity studies are just tagged on at the end. If this scheme is used, one must confront validity right from the start, as an integral part of the test. If this was always done, we would face far fewer brouhahas about the legitimacy of test usage.

Questions

1. What kinds of applications can a test scoring model that uses as its fundamental unit testlets that are built to be parallel?
2. How can a polytomous IRT model be used to model a testlet-based test?
3. What is validity-based scoring? How can it help us prevent inappropriate uses of tests?

References

Bock, R. D. (1972). Estimating item parameters and latent ability when responses are scored in two or more latent categories. *Psychometrika, 37*, 29–51.

Bock, R. D., Gibbons, R., & Muraki, E. (1988). Full information item factor analysis. *Applied Psychological Measurement, 12*, 261–280.

Kingston, N. (1987). *Feasibility of using IRT-based methods for Divisions D, E and I of the Architect Registration Examination.* Report prepared for the National Council of Architectural Registration Boards. Princeton, NJ: Educational Testing Service.

Lewis, C. (in preparation). *Validity-based scoring.* Manuscript in preparation, Educational Testing Service, Princeton, NJ.

Lewis, C., & Sheehan, K. (1990). Using Bayesian decision theory to design a computerized mastery test. *Applied Psychological Measurement, 14*, 367–386.

Lord, F. M., & Novick, M. (1968). *Statistical theories of mental test scores.* Reading, MA: Addison-Wesley.

Rosenbaum, P. R. (1988). A note on item bundles. *Psychometrika, 53*, 349–360.

Samejima, F. (1969). Estimation of latent ability using a response pattern of graded scores (*Psychometric Monograph No. 17*). Richmond, VA: Psychometric Society.

Thissen, D., Steinberg, L., & Mooney, J. A. (1989). Trace lines for testlets: A use of multiple-categorical-response models. *Journal of Educational Measurement, 26*, 247–260.

Wainer, H., & Kiely, G. L. (1987). Item clusters and computerized adaptive testing: A case for testlets. *Journal of Educational Measurement, 24*, 185–201.

Wald, A. (1947). *Sequential analysis*, New York: Wiley.

6

Fitting testlets with polytomous IRT models

6.1. Introduction

In Chapter 5, we described three different approaches to the scoring of tests made up of testlets. The aim of that chapter was to illustrate that blindly choosing an off-the-shelf test scoring model may not always be the wisest way to proceed. It is sensible to first consider the goals of the testing program before deciding how to score – one must understand what is the question being asked before trying to figure out what is the best answer. Nevertheless, we have found that a latent trait approach to test scoring is almost always useful and often may be close to optimal. There are two broad avenues that have been frequently used to solve practical testing problems. In this chapter, we will explore one of them, and the rest of this book will be devoted to the other.

The first latent trait approach leans heavily on Rosenbaum's theorem of item bundles using a polytomous item response theory (IRT) model to score the locally independent bundles (testlets). The key idea is that the items that form each testlet may have excessive local dependence, but that once the entire testlet is considered as a single unit and scored polytomously (e.g., if it is a seven-item testlet, the examinee's score for that testlet can range from 0 to 7), these local dependencies may disappear. If the testlets themselves are conditionally independent, they can be scored with a traditional polytomous IRT model.

This approach can immediately take advantage of all of the well-known characteristics of IRT (Samejima, 1969; Lord, 1980) as well as the widely distributed software to implement it (e.g., MULTILOG – Thissen, 1991) that yields:

1. interpretable item parameters,
2. their standard errors, and
3. a measure of goodness of fit.

In addition to these familiar benefits are the well-known functions of the parameters that can be used to yield expected score functions, which can be considered a testlet analog to the item characteristic curve (ICC), and testlet information functions, which show explicitly the contribution of the testlet to the total test.

In this chapter, we demonstrate, through an extended example, how this approach can yield a full and useful analysis of operational test data. The data we chose for this are from the Law School Admissions Test (LSAT).

The LSAT is composed of four sections. There is one section each of reading comprehension and analytic reasoning questions and two sections of logical reasoning. Each logical reasoning section is made up of about 25 independent items[1] and as such their statistical characteristics can be evaluated with standard procedures (e.g., Gulliksen, 1950, 1987). Reading comprehension and analytic reasoning are not made up of independent items. In both cases, their items are clustered around common stems. In reading comprehension, there are 28 items that address one of four reading passages. In analytic reasoning, there are 24 items related to one of four situations. In no cases are there fewer than five or more than eight items attached together through a common stem. The common stem that these items shared provides strong *prima facie* evidence for the invalidity of the assumption of local conditional independence required for traditional analyses.

We reach this conclusion from observing that while the assumption of local independence is an apt description for many tests composed of small, discrete items, there are other tests on which the items do not appear to be locally independent. Yen (1992) cataloged a large number of possible causes of local dependence; for instance, the several questions following each of a few reading comprehension passages typically exhibit local dependence, or the items at the end of a long (or somewhat speeded) test may similarly covary more strongly than would be predicted by an IRT model.

Items exhibiting local dependence are, to some extent, redundant. To the extent that the response to the second item in a pair depends on (and, therefore, can be predicted from) the response to the first item, beyond prediction from the underlying proficiency being measured, the second item provides less information about proficiency than would a completely (locally) independent item. Nevertheless, there are many reasons to include locally dependent groups

[1] This is almost accurate, but not quite. In one section, there were 20 independent items and two pairs of items. In the other section, there was 21 independent items and two pairs of items. For all practical purposes, one can think of these sections as having 24 and 25 independent items, respectively. Our analyses modeled the pairing, but the results are sufficiently similar to those obtained by ignoring it, which we felt that such fastidiousness was unnecessary for any practical purpose.

of items on tests. Passages followed by questions are often thought to be the most authentic of reading comprehension tests, although the cluster of questions after each passage often exhibits local dependence. Thus, the concept of the *testlet* would seem appropriate. By applying multiple-categorical IRT models to responses of conditionally independent testlets, instead of to responses to the locally dependent items within the testlets, we have found that the statistics of the IRT model generally perform better (see Sireci, Thissen, & Wainer, 1991; Thissen, 1993; Thissen & Steinberg, 1988; Thissen, Steinberg, & Mooney, 1989; Wainer & Lewis, 1990; Wainer, Sireci, & Thissen, 1991).

This chapter has, as its goal, the careful examination of the LSAT through the use of these methods to model its inherent, locally dependent structure. We will focus our attention on two of the most important aspects of the test: its precision (as measured by its reliability) and its fairness (as measured by the comparability of its performance across all of the identified subgroups of examinees taking the exam). The explicit modeling of the testlet structure of the LSAT has no drawbacks other than a slightly increased complexity of calculation, but has important advantages. If we are incorrect in our modeling, that is, if the testlets that we believe to be locally dependent turn out to be independent, our results should mirror those obtained through traditional methods. If our assumptions about the structure are correct, the statistical characterization that emerges is a more accurate one.[2]

In section 6.2, we review the historical background and methodology associated with how reliability is measured traditionally and under a testlet-based formulation. In section 6.3, we describe the methodology surrounding testlet-based DIF analysis. Section 6.4 provides the results of these analyses on the two forms of the LSAT examined, and section 6.5 contains conclusions, discussion, implications, and some suggestions for future work.

6.2. Reliability

The fact that the reliability of tests built from testlets would be overestimated by item-based methods is not newly discovered. It has been well-known for at least 70 years. Warnings about inflated reliability estimates are commonly carried in many introductory measurement textbooks. For example,

> One precaution to be observed in making such an odd-even split pertains to groups of items dealing with a single problem, such as questions referring to a particular mechanical diagram or to a given passage in a reading test. In this case, a whole

[2] Seeking accuracy from analytic answers is far from new, although some may think it so. For example, 1,700 years before Samuel Johnson's observation (quoted in Chapter 1), Seneca was well aware of the difference – *"Magnum esse solem philosophus probabit, quantus sit mathematicus."*

> group of items should be assigned intact to one or the other half. Were the items in such a group to be placed in different halves of the test, the similarity of the half scores would be spuriously inflated, since any single error in understanding of the problem might affect items in both halves.
>
> *Anastasi (1961, p. 121; included in all subsequent editions as well, see 6th edition, p. 121)*

> In some cases, several items may be unduly closely related in content. Examples of this would be a group of reading comprehension items all based on the same passage.... In that case, it will be preferable to put all items in a single group into a single half-score... this procedure may be expected to give a somewhat more conservative and a more appropriate estimate of reliability....
>
> *Thorndike (1951, p. 585)*

> Interdependent items tend to reduce the reliability. Such items are passed or failed together and this has the equivalent result of reducing the length of the test.
>
> *Guilford (1936, p. 417)*

These warnings were generated by an earlier exchange about the potential usefulness of the Spearman–Brown prophecy formula. This exchange is summarized by Brown and Thomson (1925), who reported criticisms of the mathematician W. L. Crumm (1923). Crumm felt that the requirements of Spearman–Brown were too stringent to expect them to be met in practice. Holzinger (1923) reported some empirical evidence supporting the Spearman–Brown formula, and then Kelley (1924) provided a fuller mathematical defense of Spearman–Brown. Of interest here is Kelley's (1924) comment, "If two or more exercises contain common features, not found in the general field, then the Spearman–Brown r_{1I} will tend on this account to be too large" (p. 195).

All of these are in the context of even–odd split-half reliability; but coefficient α averages all of the split-half reliabilities. Thus, if some of the split-half reliabilities are inflated by within-passage correlation, so too would be α.

The APA *Standards for Educational and Psychological Tests* (1966, 1974, 1985, Section D5.4) deemed treating testlets in a unitary matter "essential." It stated

> If several questions within a test are experimentally linked so that the reaction to one question influences the reaction to another, the entire group of questions should be treated preferably as an "item" when the data arising from application of split-half or appropriate analysis-of-variance methods are reported in the test manual.

This led to a number of researchers (Wainer & Kiely, 1987; Thissen, Steinberg, & Mooney, 1989; Wainer & Lewis, 1990; Wainer, Sireci, & Thissen, 1991) to propose methods for treating testlets as the fundamental building blocks of tests; here we concentrate on how the estimates of reliability might be affected as a result of failing to do so. Next, after we briefly review what was previously discussed in Chapters 2 and 3, we shall expand a bit on the improvement in accuracy yielded by taking this approach.

6.2.1. Methodology

Reliability is defined (Lord & Novick, 1968, p. 61) as the squared correlation $\rho^2_{x\tau}$ between observed score (x) and true score (τ). This could be expressed as the ratio of true score variance to observed score variance, or after a little algebra

$$\rho^2_{x\tau} = \frac{\sigma^2_x - \sigma^2_e}{\sigma^2_x}, \tag{6.1}$$

where the subscript e denotes error, as described in Chapter 2. In this chapter, we use coefficient α (see Lord & Novick, 1968, p. 204) to estimate the reliability of the summed scores of the traditional test theory.

There are many ways to calculate reliability, but in traditional test theory it takes a single value that describes the average error variance for all scores. (Traub & Rowley, 1991, p. 40, use the notation Ave(σ^2_e) in place of σ^2_e to emphasize that only an average estimate of the error variance is considered in true score reliability theory.) In contrast to this simplification, measurement precision in an IRT system can be characterized as a function of proficiency (θ); therefore, precision need not be represented by a single overall "reliability." Precision in an IRT system is usually described in terms of $I(\theta)$, the information function, the conditional error variance σ^2_{e*}, or the standard error σ_{e*}; these also vary as functions of θ. Although measurement error may vary as a function of proficiency, Green et al. (1984) observed that it could also be averaged to give "marginal reliability" comparable to that of the traditional theory. The marginal measurement error variance, $\bar{\sigma}^2_{e*}$, for a population with proficiency density $g(\theta)$ is

$$\bar{\sigma}^2_{e*} = \int \sigma^2_{e*} g(\theta) d\theta, \tag{6.2}$$

where σ^2_{e*} is the expected value of the error variance associated with the *expected a posteriori* estimate at θ, and the marginal reliability is

$$\bar{\rho} = \frac{\sigma^2_\theta - \bar{\sigma}^2_{e*}}{\sigma^2_\theta}. \tag{6.3}$$

Here, the integration (or averaging) over possible values of θ takes the place of the traditional characterization of an average error variance. The value of $\bar{\sigma}^2_{e*}$ is the average of the (possibly varying) values of the expected error variance, σ^2_{e*}; if many values of σ^2_{e*} were tabulated in a row for all the different values of proficiency (θ), $\bar{\sigma}^2_{e*}$ would be that row's weighted marginal average. Therefore, the reliability derived from that marginal error variance is called the *marginal* *reliability*, and denoted by $\bar{r}$ to indicate explicitly that it is an average. There is some loss of information in averaging unequal values of σ_{e*}^2. But such marginal reliabilities for IRT scores parallel the construction of internal consistency estimates of reliability for traditional test scores.[3] In the remainder of this chapter, we use $\bar{\rho}$ to describe the reliability of estimates of latent proficiency (θ) based on IRT.

The surface analogy between Equations (6.1) and (6.3) is slightly deceiving. The term σ_{e*}^2 is the expected value of the error variance of the *expected a posteriori* estimate of θ as a function of θ, computed from the information function for the *expected a posteriori* estimate of θ – as Birnbaum (1968, pp. 472ff) illustrates, any method of test scoring has an information function; here, we use the information function for the *expected a posteriori* estimate. That is analogous to the error of estimation, σ_ε^2, in the traditional theory, not σ_e^2 (see Lord & Novick, 1968, p. 67). And σ_θ^2 is the true population value of the variance of θ.

The definition of traditional reliability in terms of true score variance σ_T^2 and the error of estimation σ_ε^2 may be straightforwardly derived from Lord and Novick's (1968, p. 67) Equation (3.8.4):

$$\sigma_\varepsilon = \sigma_T\sqrt{1 - \rho_{XT}^2},$$

which gives

$$\rho_{XT}^2 = \frac{\sigma_T^2 - \sigma_\varepsilon^2}{\sigma_T^2};$$

that is, the classical analog of Equation (6.3).

6.2.2. An IRT model for testlets

Although it is possible to approach testlet analysis (the item analysis of testlets) using the tools of the traditional test theory, we have found the tools of IRT to be more useful. In the computation of the IRT index of reliability, $\bar{\rho}$, we follow Thissen, Steinberg, and Mooney (1989) in their use of the specialization

[3] Technically, Mislevy (personal communication, July 1990) has pointed out that if the information function is steeper than the proficiency distribution, the integral in (6.2) can become unbounded. This is typically not a problem in applications where one approximates the integral with a sum. Yet, one could imagine a case in which changing the bounds of the integral could drastically alter results. Mislevy's alternative is to integrate the information function directly and then invert. Specifically, substitute for (6.2) the expression

$$\bar{I}(\theta) = \int_{-\infty}^{\infty} I(\theta)g(\theta)\,d\theta = \frac{1}{\sigma_{e*}^2}. \tag{6.2*}$$

Mislevy's formulation has obvious theoretical advantages, but we have too little experience as yet to indicate under what circumstances the extra protection is needed.

of Bock's (1972) model introduced in Chapter 5. To recapitulate, we have J testlets, indexed by j, where $j = 1, 2, \ldots, J$. On each testlet, there are m_j questions, so that for the jth testlet there is the possibility for the polytomous response $x_j = 0, 1, 2, \ldots, m_j$.The statistical testlet scoring model posits a single underlying (and unobserved) dimension, which we call latent proficiency, and denote as θ. The model then represents the probability of obtaining any particular score as a function of proficiency. For each testlet, there is a set of functions, one for each response category. These functions are sometimes called item response functions (IRFs) (Holland, 1990).

The IRF for score $x = 0, 1, \ldots, m_j$, for testlet j is, as defined in Chapter 5,

$$P(T_{jx} = m_j) = \frac{\exp[a_{jk}\theta + c_{jk}]}{\sum_{k=0}^{m_j} \exp[a_{jk}\theta + c_{jk}]}, \tag{6.4}$$

where $\{a_k, c_k\}_j$, $k = 0, 1, \ldots, m_j$ are the item category parameters that characterize the shape of the individual response functions. If this model yields a satisfactory fit, information is calculated and inverted to yield estimates of the error variance function. Error variance is relative to the variance of θ whose distribution $g(\theta)$ is fixed as $N(0,1)$ to help identify the model. The integration indicated in (6.2) could then be carried out and $\bar{\rho}$ calculated through Equation (6.3). This model is discussed in greater detail in the next section.

6.3. Testlet-based DIF

The past decade or so has seen a renewed emphasis on issues of test fairness. One aspect of fairness is the insistence that test items do not function differentially for individuals of the same proficiency, regardless of their group membership. "No DIFferential item functioning" is now a general desideratum; the area of study surrounding this has been defined formally and dubbed as DIF (see Holland & Wainer, 1993, for a detailed description of this entire area). A set of statistically rigorous and efficient procedures have been developed to detect and measure DIF. These generally fall into one of two classes; they are based either on latent variables (Thissen, Steinberg, & Wainer, 1988, 1993) or on observed score (Holland & Thayer, 1988; Dorans & Holland, 1993); there are also a variety of methods that fall in-between (Swaminathan & Rogers, 1990; Shealy & Stout, 1993)

Procedures for DIF studies have traditionally focused on the item. Yet, if a test is based on a broader unit (testlet), ought we not use a generalized DIF procedure that suits this broader construct? The aim of this section is to show

how to utilize such a generalization and to provide estimates of the extent to which the LSAT conforms to contemporary standards of fairness.

The determination of DIF at the testlet level has three advantages over confining the investigation to the item. It allows:

1. the analysis model to match the test construction;
2. DIF cancellation through balancing; and
3. the uncovering of DIF that, because of its size, evades detection at the item level but can become visible in the aggregate.

6.3.1. Matching the model to the test

If a set of items were built to be administered as a unit, it is important that they be analyzed that way. There are a variety of reasons for analyzing them as a unit, but underlying them all is that if one does not, one is likely to get the wrong answer. In an example described earlier (Wainer, Sireci, & Thissen, 1991), a four-testlet test consisting of 45 separate items yielded a reliability of 0.87 if calculated using traditional methods assuming the 45 items were conditionally independent. If one calculated reliability taking the excess within-testlet dependencies into account, the test's reliability is shown to be 0.76. These are quite different – note that Spearman–Brown (Gulliksen, 1950/1987, p. 78) indicates that we would need to double the test's length to yield such a gain in reliability (see Sireci, Thissen, & Wainer, 1991, for more details on this aspect). Other calculations (i.e., validity and information) are affected as well.

6.3.2. DIF cancellation

Roznowski (1988), among others, has pointed out that because decisions are made at the scale or test level, DIF at the item level may have only limited importance. Therefore, it is sensible to consider an aggregate measure of DIF. Small amounts of item DIF that cancel within the testlet would seem, under this argument, to yield a perfectly acceptable test construction unit.

Humphreys (1962, 1970, 1981, 1986) has long argued that it is both inadvisable and difficult – very likely impossible – to try to construct a test of strictly unidimensional items. He suggests that to do so would be to construct a test that is sterile and too far abstracted from what would be commonly encountered to be worthwhile. He recommends the use of content-rich (i.e., possibly multidimensional) items, and since multidimensionality is one cause of DIF, we ought to control it by balancing across items. We agree with this. But balancing is not a trivial task. Surely such balancing needs to be done within the content area and across the entire test. For example, it would be unfortunate if the items that favored one group were all present at the end of the test. The concept of a testlet suggests itself naturally. Build the test out of testlets and ensure that there is no DIF at the testlet level. Lewis and Sheehan (1990) have shown that building a mastery test of parallel-form testlets provides a graceful solution to a set of thorny problems.

A final argument in support of examining DIF at the testlet level derives from the consideration of testlets that cannot be easily decomposed into items. For example, consider a multistep mathematics problem in which students get credit for each part successfully completed. Does it make sense to say that parts of such a testlet contain "positive subtraction DIF" and then "negative multiplication DIF"? Of course not. Instead, we must concentrate on the DIF of the problem as a whole. In some sense, we do this now with dichotomously scored multiple-choice items when we test an item's DIF. We do not record intermediate results and so do not know to what extent there is DIF on the component tasks required to complete the item. All we concern ourselves with is the final result.

It should be emphasized that by cancel out we mean something quite specific. We mean that there will be no DIF at *every* ability level within the testlet. Exactly how we operationalize this goal and what it means will be explicated and illustrated in the next sections.

6.3.3. Increased sensitivity of detection

It is possible (and perhaps sometimes even likely) to construct a testlet of items with no detectable item DIF, yet the testlet in the aggregate does have DIF. The increased statistical power of dealing with DIF at the testlet level provides us with another tool to ensure fairness. This can be especially useful for those focal groups that are relatively rare in the examinee population and so are not likely to provide large samples during item pretesting. As will be shown in the results section, this was the case in the operational forms of the LSAT we analyzed.

6.4. Methodology

6.4.1. Testlet DIF detection

The polytomous IRT model we used (specified in Equation (6.4)), was developed by Bock (1972). For the detection of testlet DIF, we use the well-developed technology of likelihood ratio tests. The properties of these tests have been thoroughly investigated (Kendall & Stuart, 1967). Of primary importance in this work is their near optimality. Under reasonable conditions, these tests are closely related to, although not necessarily identical to, the most powerful test given by the Neyman–Pearson lemma (Neyman & Person, 1928). This optimality of power is critical in situations such as DIF detection, in which the testing organization would like to accept the null hypothesis; look for DIF and not find any. Not finding DIF is easily done by running poor experiments with weak statistical methods. Thus, to be credible one must use the largest samples available as well as the most powerful analytic tools. In this investigation, we used all of the examinees and the likelihood ratio test. Current statistical standards for DIF detection (Holland & Wainer, 1993) acknowledge that this is the best that can be done at the moment.

The basic notion of the likelihood ratio test is to fit the model to the data, assuming that all testlets have the same parameters (no DIF) in the two populations of interest (*reference* and *focal*). Next, fit the same model to the data, allowing one testlet to have different parameters in each population (DIF) and compare the likelihood under each of the two situations. If the more general model does not yield a significant increase in the quality of the fit, we conclude that the extra generality was not needed and that the testlet in question has no DIF. This procedure was applied in the study of DIF by Thissen, Steinberg, and Wainer (1988) using a more traditional dichotomous IRT model. Thissen et al. used Bock's polytomous model to fit testlets (Thissen, Steinberg, & Fitzpatrick, 1989; Thissen, Steinberg, Mooney, 1989). Our testlet approach to DIF is almost exactly the one reported by Thissen, Steinberg, and Wainer (1993) when we used the multiple-choice model (Thissen & Steinberg, 1984) to examine differential alternative functioning (DAF). The step from DAF to testlet DIF is a small one.

Bock's 1972 model

As described in Chapter 5, we can represent the trace line for score $x = 0, 1, \ldots, m_j$, for testlet j, as

$$P(T_{jx} = m_j) = \frac{\exp[a_{jk}\,\theta + c_{jk}]}{\sum_{k=0}^{m_j} \exp[a_{jk}\,\theta + c_{jk}]}.$$

The a_k's are analogous to discriminations; the c_k's analogous to intercepts. The model is not fully identified, and so we need to impose some additional constraints. It is convenient to insist that the sum of each of the sets of parameters equals zero, that is,

$$\sum_{k=0}^{m_j} a_{jk} = \sum_{k=0}^{m_j} c_{jk} = 0.$$

In this context, we reparameterize the model using centered polynomials of the associated scores to represent the category-to-category change in the a_k's and the c_k's:

$$a_{jk} = \sum_{p=1}^{P} \alpha_{jkp} \left(k - \frac{m_j}{2}\right)^p \tag{6.5}$$

and

$$c_{jk} = \sum_{p=1}^{P} \gamma_{jp} \left(k - \frac{m_j}{2}\right)^p, \tag{6.6}$$

where the parameters $\{\alpha_p, \gamma_p\}_j$, $p = 1, 2, \ldots, P$, for $P \leq m_j$ are the free parameters to be estimated from the data. The polynomial representation has, in the past, saved degrees of freedom with no significant loss of accuracy. It also provides a check on the fit of the model when the categories are ordered. Although this model was developed for the nominal case, it can also be used for ordered categories. If the a's are ordered, the categories must be monotonically ordered as well (see Wainer, Sireci, & Thissen, 1991, for proof). This specialization of Bock's nominal model is often referred to as Samejima's (1969) graded response (partial credit) model. The polynomial representation in this application saves degrees of freedom while usually providing a good representation of the data.

This version of Bock's model uses raw score within testlet as the carrier of information. While it is possible that more information might be obtained by taking into account the pattern of responses within each testlet, we found that this simplification is appropriate. Moreover, basing a test-scoring algorithm on number right is amply supported by general practice.

In Chapter 5, we pointed out that this model was fitted to a four-passage, 22-item test of reading comprehension by Thissen, Steinberg, and Mooney (1989), who found that there was a slightly greater validity of the testlet-derived scores when correlated with an external criterion, and that the test information function yielded by the traditional analysis was much too high. The testlet approach thus provided a more accurate estimate of the accuracy of the assessment, and through the obvious generalization, this same approach can be used to study testlet DIF.

The basic data matrix of score patterns is shown in Table 6.1. In this example, there are four testlets with 10 possible scores levels each [$m_j = (10, 10, 10, 10)$]; there are a maximum of 10^4 rows. In practice, there would be far fewer rows since many possible response patterns would not appear. The analysis follows what is done in item DIF situations: fitting one model, allowing different values for the parameters of the studied testlet for the two groups and then comparing the -2loglikelihoods of that model with others that restrict the two groups'

Table 6.1. *Arrangement of the data for the IRT analyses*

Testlet score pattern					Frequencies	
I	II	III	IV	Total Score	Reference	Focal
0	0	0	0	0	f_{R1}	f_{F1}
0	0	0	1	1	f_{R2}	f_{F2}
0	0	0	2	2	f_{R3}	f_{F3}
.	.	.	.	.	.	.
.	.	.	.	.	.	.
.	.	.	.	.	.	.
.	.	.	.	.	.	.
s_I	s_{II}	s_{III}	s_{IV}	$\sum s_j$	f_{Ri}	f_{Fi}
.	.	.	.	.	.	.
.	.	.	.	.	.	.
.	.	.	.	.	.	.
9	9	9	9	36	f_{RN}	f_{FN}

estimates in a variety of ways. Stratification/conditioning is done on θ estimated for both groups simultaneously.

This method uses the test itself, including the studied testlet, to calculate the matching criterion. The question about whether or not to include the studied item has been carefully explored by Holland and Thayer (1988) who showed for the Rasch model (the binary analog of this model) that *not* including the studied item in the criterion yields statistical bias under the null hypothesis. This was explored further by Zwick (1990) who confirmed this result for the Rasch model, but not generally for other IRT models.

Using this method requires first fitting a completely unrestricted model – estimating all of the a_k's and c_k's separately for both the reference and the focal groups. Next restricted versions of this model are estimated by approximating the values of the parameters as polynomial functions of score category (Equations (6.5) and (6.6)). When an acceptably fitting parsimonious model is derived, we note the value of -2loglikelihood (asymptotically χ^2) for that model and then sequentially restrict the parameters for one testlet at a time to be equal across the two groups. We subtract the value of -2loglikelihood of the restricted model from that of the unrestricted one, and, remembering that if we subtract a smaller χ^2 from a larger one, the difference between the two is also χ^2. We test that difference for significance: the number of degrees of freedom of the statistical test is equal to the number of parameters restricted. If it is not significant, we conclude that the extra flexibility gained by allowing different parameters for the focal and reference groups is not required – there is no DIF. If it is significant, we can further isolate where the DIF is located.

Eventually, one arrives at a determination of the most parsimonious representation. Interpreting the character of this representation allows us to detect testlet DIF. This is computationally expensive, with the cost of each run essentially linear in the number of response patterns observed. Of course, this cost is small relative to the cost of not detecting testlet DIF when it is there. The cost can be controlled substantially by reducing the number of possible response patterns.

The method of analysis utilized here departs from complex IRT analyses of the past in an important way. There is no *post hoc* "linking" of analyses through common items and *ad hoc* assumptions about stability and linearity. Instead, we specify the shape and location of the proficiency distribution. This not only provides a common metric for all comparisons (the trait in the population) but also yields estimates of the correct likelihood, thus allowing inferences about the significance of discovered effects to stand on firmer statistical ground. This methodology is based on the estimation procedure of maximum marginal likelihood[4] (MML) developed by Bock and Aitken (1981). Thus, within each analysis, we estimate the item parameters as well as the distribution of the proficiency distribution of the examinees. If there is just a single group, the proficiency distribution is set at $N(0, 1)$, but when there is more than a single group (as in the DIF analyses), we set one group's proficiency distribution at $N(0, 1)$ and estimate the mean of the other groups' distribution. Because this automatically adjusts for differences in the overall proficiencies of the various subgroups, we can maintain all item parameters on the same scale. Moreover, we can then compare the overall performance of the various subgroups in a concise and meaningful way. We will report such a comparison among the various examinee subgroups on the LSAT as the initial part of our results section.

6.5. Results

All analyses described above were performed on all four sections of two parallel forms of the LSAT. The polytomous IRT model fit well on each section. Its fit deteriorated significantly when dissimilar sections were combined. If this finding is replicated on other forms, it strongly suggests that inferences made

[4] MML was called "marginal maximum likelihood" by Bock and Aitken when they invented it. But they were wrong. The method they proposed does not marginalize the maximum likelihood estimate; it maximizes the marginal likelihood. Thus, this restatement of what MML means more accurately describes the method. Bock agrees and has started calling it maximum marginal likelihood too (despite this permutation in the name, it still remains appropriate to call it "Mummel" = MML).

Table 6.2. *Proficiency distribution means*

	Reading comprehension	Analytic reasoning	Logical reasoning (*a*)	Logical reasoning (*b*)	**Overall**
Male	0.10	0.00	0.26	0.37	**0.18**
Female	0.00	0.00	0.00	0.00	**0.00**
White	0.00	0.00	0.00	0.00	**0.00**
African-American	−0.65	−0.92	−1.16	−1.18	**−0.98**
Hispanic	−0.40	−0.51	−0.69	−0.74	**−0.59**
Asian	−0.32	−0.19	−0.19	−0.37	**−0.27**
U.S.	0.27	0.09	−0.04	0.02	**0.09**
Canadian	0.00	0.00	0.00	0.00	**0.00**

from fitting a unidimensional model to the LSAT in its entirety should be limited as much as possible, and avoided in many instances. Happily, our purposes did not require such aggregated fitting.

Our findings can be divided into three categories:

1. *overall performance* of individual subgroups on the test,
2. the *reliability* of each section, and
3. the *differential performance* of items/testlets.

6.5.1. Overall performance

Each section of the test was scaled separately within the various subanalyses. One group was assigned an arbitrary mean level of performance of 0.0 and the other group's performance was characterized relative to that. All figures are in standard deviation units; the distribution of scores within groups can be thought of as Gaussian. Thus, we see that men scored about 0.1 above women in reading comprehension, were essentially identical to women in analytic reasoning, and were about a third of a standard deviation higher in logical reasoning (Table 6.2). On average (weighting all sections equally), men scored about two tenths of a standard deviation higher than did women.

Obviously, any recommendations made on the basis of the findings from the analyses of only these two LSAT forms (which are really the same form with the presentation order of the sections shuffled) must be confirmed on other forms before being considered for implementation. These results are suggestive (albeit sometimes strongly so), not conclusive.

Table 6.3. *Group relibilities*

	Reading comprehension	Analytic reasoning	Logical reasoning (a)	Logical reasoning (b)
Male	0.69	0.60	0.79	0.77
Female	0.69	0.60	0.77	0.77
White	0.69	0.58	0.78	0.75
African-American	0.68	0.56	0.69	0.79
Hispanic	0.69	0.60	0.75	0.72
Asian	0.71	0.59	0.77	0.77
U.S.	0.68	0.59	0.79	0.77
Canadian	0.68	0.59	0.79	0.77
Overall	**0.71**	**0.60**	**0.79**	**0.79**

Table 6.4. *Reliability of the LSAT*

Section	Item-based reliability	Testlet-based reliability	Reliability range	Number of items	Length increase required
Reading comprehension	0.79	0.71	0.68–0.71	28	1.54
Analytic reasoning	0.76	0.60	0.56–0.60	24	2.11
Logical reasoning (a)	0.77	0.79	0.69–0.79	24	0.89
Logical reasoning (b)	0.77	0.79	0.72–0.79	25	0.89

6.5.2. Reliability

Shown in Table 6.3 are the reliabilities of each section of the LSAT within the various subgroups specified. As can be seen, there is very little variation in section reliability across subgroup and so one can productively use the single overall value shown at the bottom. What differences there are in the section reliability among groups is largest in the logical reasoning section.

Shown in Table 6.4 is a comparison of these reliabilities with those obtained using traditional methodology (labeled "Item-based reliability" in the table) that does not take into account the testlet structure of the LSAT.

The column labeled "Reliability range" reflects the variability of reliability seen across all subgroups (reported explicitly in the table that immediately preceded this one). The column labeled "Length increase required" is an estimate, obtained from the Spearman–Brown prophecy formula, of how much longer that section would have to be made in order for it to have a testlet-based estimate of reliability that is equal to the estimate obtained through traditional test theory when the latter does not explicitly model the local dependence observed. Analytic reasoning would have to be more than doubled, yielding eight situations instead of the current four. Similarly, reading comprehension would require six passages rather than the current four to yield the reliability of 0.79 that was previously claimed.

These findings may have implications for the way that the various test sections are combined to yield total score. If all sections are equally reliable, there are psychometric arguments in support of counting each section equally. The nominal reliability of the four sections examined are sufficiently close to warrant this approach. However, if we accept the testlet-based estimates of reliability, all sections are no longer equally reliable. If we want the composite score to be maximally reliable, we must either weight the various sections differentially (Wang & Stanley, 1970) or make the various component sections equally reliable. The latter can be accomplished in at least two ways: adjusting the length of the sections or using ancillary information in the estimation of each section's score.

There are two observations that can be made from these results. First, that explicitly modeling the testlet structure of the test provides us with a more accurate estimate of the parallel forms reliability of each section. This results in a decrease in those two sections of the test that are substantially clustered. Second, the reliability of the logical reasoning sections, which does not have much clustering (there are two pairs of items), has actually increased a little bit. This increase is due to the use of IRT, which extracts more information from the test response pattern of each individual examinee than does using merely the total number correct.

6.5.3. Differential performance

Using the likelihood ratio to detect differential performance (DIF) involves heavy computation for tests of even modest size. This is especially true if some sort of purification of the stratifying variable is done – that is, detecting items/testlets that might show DIF and excluding them from the portion of the test that is used to stratify the examinees. One practical way to ease this burden is to do some preliminary exploration using the less general, but far simpler, DIF measure derived from the Mantel–Haenszel (MH) statistic. This approach allows us to easily detect items/testlets that seem to show DIF, eliminate them from the stratifying variable, and then use the more powerful likelihood ratio procedures to examine the details. This approach has shown itself to be a practical solution to a difficult problem, and was the approach we used. We also include the results of the MH analyses to illustrate how "no DIF" at the item level does not mean "no DIF" at the testlet level.

All of the individual items were screened for DIF using MH methods during pro forma item analysis. Very few were found to deviate from acceptable standards of fairness. However, when these individual items were viewed from the point of view of a testlet, a different pattern emerged. As an example of what we found, consider the MH statistics associated with the comparisons between Whites and African-Americans (below). The standard interpretation of the MH statistic (see Dorans & Holland, 1993, p. 42) classifies items as problematic if $|MH| > 1.5$. Under this rule, only one item (#3) is questionable. The sign of the statistic indicates which group is being disadvantaged; a positive value means it favors the focal group (African-Americans in this instance), a negative value the reference group (Whites). It would not be surprising to find that these two passages and their associated items passed muster after pretest screening (in those cases where there were enough data to allow it), since only one item stuck out over the statistical barrier against discrimination, and that one only barely. However, if we look at these statistics as a group, clustered by passage (Table 6.5), there is an obvious pattern. Virtually, all of the items associated with Passage 1 favor the focal group (total $MH = 4.82$) and five of the seven items associated with the fourth passage favor the reference group (total $MH = -1.38$).

The results obtained from the traditional DIF analysis were suggestive. When the data were analyzed for DIF with the testlet structure explicitly modeled, we found that at all levels of proficiency, African-Americans had a higher expected true score on this testlet than did Whites. The extent to which this is true is shown in Figure 6.1. On the other hand, Whites are advantaged on Passage 4 (Figure 6.2).

Thus, using a test scoring procedure that explicitly models the structure of the LSAT, we were able to detect and measure possible deviations from fairness that were hidden previously. Our principal substantive finding, after examining DIF for all testlets on five focal groups (women, African-Americans, Hispanics, Asian-Americans, and Canadians), is that there was very little DIF associated with any particular testlet that was consistent across most of the focal groups, other than the two shown above. Both Passages 1 and 4 of the reading comprehension section were found to consistently show differential performance.

We have used a simplifying display method, which, though not particularly arcane, may be unfamiliar. Thus, we make a brief aside to comment on it.

Table 6.5. *Mantel–Haenszel statistics shown for items grouped by passage*

Passage	Item number	Mantel–Haenszel statistic
1	1	0.88
	2	−0.04
	3	1.59
	4	0.98
	5	0.53
	6	0.88
	Total	**4.82**
4	22	0.26
	23	−0.57
	24	−0.42
	25	−0.31
	26	−0.54
	27	0.56
	28	−0.36
	Total	**−1.38**

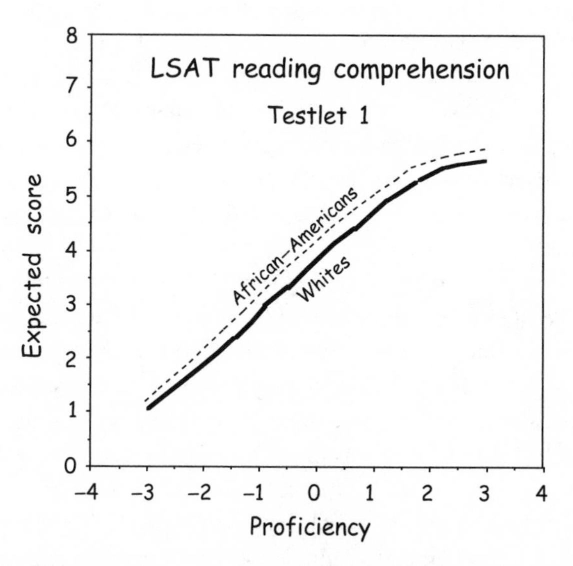

Figure 6.1. The expected true score curves for Testlet 1 for both African-American and White examinees, indicating a slight advantage for African-Americans.

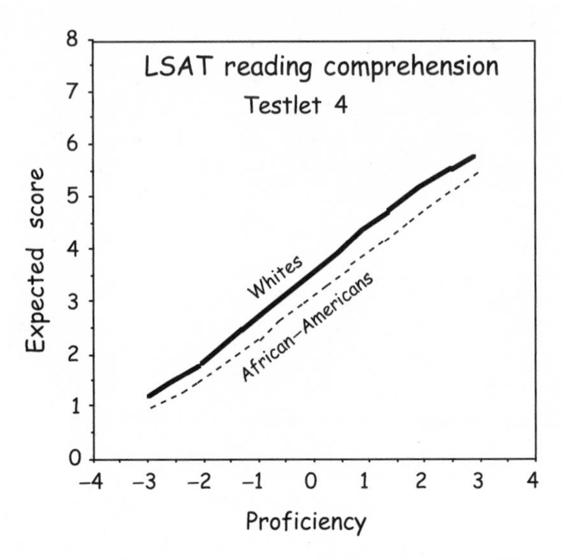

Figure 6.2. The expected true score curves for Testlet 4 for both African-American and White examinees, indicating a slight advantage for White examinees.

Dichotomous IRT traditionally represents an item's performance with the traceline associated with the probability of a correct response: a plot of p(correct) versus θ. There are really two tracelines for a dichotomous item. The one not plotted is associated with the probability of getting the item wrong $(1-p)$. In the dichotomous case, the two tracelines provide redundant information, and therefore, in the interests of parsimony, are rarely both presented. In the polytomous case, there are more than two tracelines. There is one traceline associated with being in each category. This is often presented as a sequence of tracelines. When the number of score categories is greater than 3 or 4, this practice usually results in a web of lines that defy easy interpretation. To ease this problem, we plot the expected true score curve. Because these curves are monotonic, they tend to be well behaved and quite regular in appearance. They are also in a metric that is meaningful to the observer. Expected true score is calculated by choosing a large number of points on the proficiency axis (say 20), and at each point, calculating the expected true score. This is done by multiplying the value of each score category by the probability of obtaining that score at that value of θ. The score category values are determined by the test developer; the probabilities are estimated from the model (Equation (6.4)). In DIF analyses, we use different values of the parameters for the focal and reference groups to estimate these probabilities. We have found that, because of trade-offs in the model, it is almost impossible to judge the severity of the DIF through examination of the absolute differences in the parameter values. But when these differences are characterized in the metric of expected true score, our understanding is improved substantially. Moreover, for this model, no DIF in the expected true score curves is a necessary and sufficient condition for no DIF in the component tracelines (Chang & Mazzeo, 1994).

6.5.4. A testlet examination of White–African-American performance on Reading Comprehension: a brief case study

We have reported the most substantial findings that have been uncovered. In this subsection, we step through one set of analyses, one test section for one focal group, to indicate how we arrived at these final results. It would be wasteful of effort to repeat this explanation for all sections and all focal groups.

Step 1: *Fit the joint data for White and African-American examinees with a single model*

The reading comprehension section of these forms of the LSAT consists of four testlets. Each testlet is composed of a passage and 6, 7, 8, and 7 items, respectively. Because of the possibility of a zero score, this means that the scores achievable on each testlet fell into 7, 8, 9, and 8 score categories, respectively. Thus (referring back to Table 6.1), each group had the possibility of 4,032 ($= 7\times 8 \times 9 \times 8$) score patterns. The initial analysis performed treated the two groups of examinees as one and estimated the mean of the proficiency distributions for those two groups as well as estimating the fit.[5] Because we treated both groups as one, we were explicitly assuming no DIF. We found that −2loglikelihood for this model was 8,224. After fitting this model, there were 33 free parameters. We found that a preliminary estimate of the mean of the proficiency distribution for White was 0.8σ higher than that for the African-Americans. Part of this difference may be due to testlet DIF. We will amend this figure as the analysis progresses.

Step 2: *Relax the equality constraints completely and fit a more general model*

In this model, we allowed the parameters for each of the testlets to be estimated separately for each of the two groups; ALL DIF run. This was done in a single run in the same way as Step 1 by releasing the equality constraints

[5] As a brief operational aside, we ran this model as if there were eight testlets, but Whites omitted the last four and African-Americans omitted the first four. Then we applied equality constraints to set Testlet 1 = Testlet 5, Testlet 2 = Testlet 6, Testlet 3 = Testlet 7, and Testlet 4 = Testlet 8. This effectively fits a single model of NO DIF while allowing separate estimates of proficiency to be obtained for each group. It also provides a graceful way to fit successively more general models.

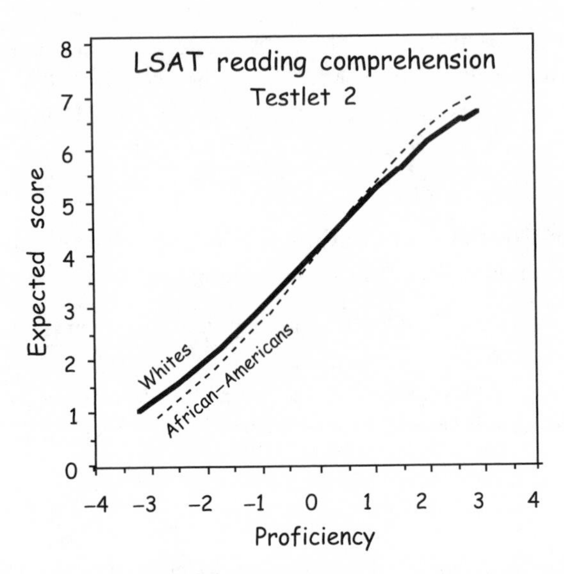

Figure 6.3. The expected true score curves for Testlet 2 for both African-American and White examinees, indicating essentially no advantage to any group.

on the parameters while fixing the proficiency distributions to be the same as those estimated in Step 1. This yielded a -2loglikelihood of 7,408 but with 64 free parameters.

Step 3: *Compare NO DIF with ALL DIF*

Subtracting the fits of the two runs obtained in Steps 1 and 2 resulted in a χ^2 of 816 (8,224 – 7,408) on 31 (64 – 33) degrees of freedom. This is statistically significant by any measure. Thus, we can reject the NO DIF model. The task was then to track down where, which testlet(s), had the DIF.

Step 4: *Exploratory analysis – plot the tracelines from the ALL DIF analysis*

From the finding in Step 3, we knew that at least one and at most four testlets had White–African-American DIF. Tracking down which one(s) is easier with some exploration, we plotted the expected true score curves for each testlet. From these, we observe that there is very little difference between the expected true score curves for Testlets 1 and 2, as compared to more substantial differences between the groups for Testlets 3 and 4 (Figures 6.3 and 6.4).

This observation led us to the next step.

Step 5: *Fit a model that sets Testlets 1 and 2 equal in the two groups, but allows Testlets 3 and 4 to be unequal.*

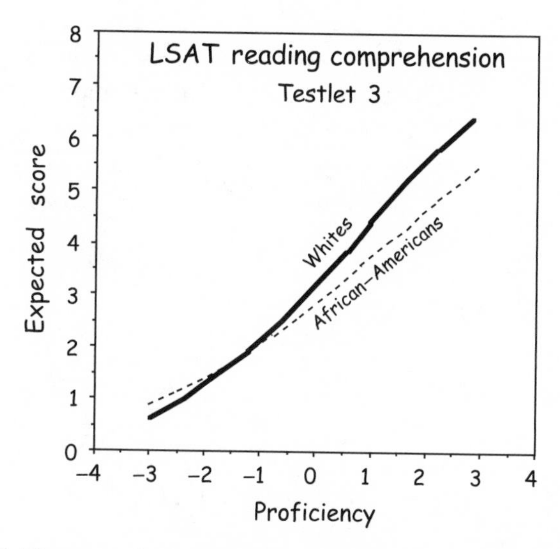

Figure 6.4. The expected true score curves for Testlet 3 for both African-American and White examinees, indicating a slight advantage for higher ability White examinees.

When this model was fit (3 & 4 DIF), we found that −2loglikelihood was 7,528 and there were 50 free parameters. This model fits much better than the NO DIF model, (χ^2 of 696 on 17 degrees of freedom) but is still significantly worse than the ALL DIF case (χ^2 of 120 on 14 degrees of freedom). Inspection of the previous figures led us to fit a model in which only Testlet 1 was fixed to be equal in the two groups.

Step 6: *Fit a model with Testlets 2, 3, & 4 unequal*

When this model was fit, the −2loglikelihood was 7,422, with 58 free parameters. This represented a significant improvement over the previous model (Step 5) with a χ^2 of 106 on 8 degrees of freedom and it was not seriously worse than the ALL DIF model (χ^2 of 14 on 6 degrees of freedom). It was this model that we chose to represent the data. This model was fit without fixing the means of the proficiency distributions, and the estimates of those means thus obtained indicted that Whites were 0.65σ higher than those for African-Americans. Note that this is a smaller difference than was observed in the original NO DIF run. The reason for this is that some of the difference between the two groups was absorbed by allowing three of the testlets to have different parameters. Thus, by adjusting for the DIF, a smaller estimate of the between-group difference was obtained.

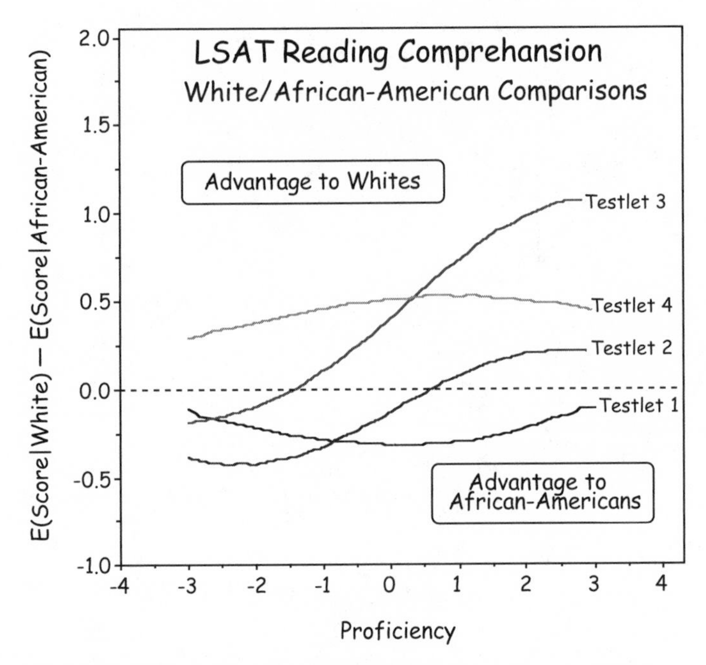

Figure 6.5. A summary of all four testlets explicitly showing the difference between the expected true score curves for White and African-American examinees as a function of proficiency. At lower levels of proficiency, there is a slight advantage for African-Americans on three of the four testlets; at high levels this is reversed.

There are two details that deserve mention. First, although the likelihood ratio comparing ALL DIF with 2, 3, & 4 DIF models is modest, compared to the other ones, it still achieves a nominal value of significance. This ought not be followed slavishly, for with large enough sample sizes any observed difference would be "statistically significant." In this study, we had a focal group of 3,149 and a reference group of 33,681. With such sample sizes, very small differences are detectable. The absolute size of the difference between the two curves is of greater importance. Note that the tiny difference observed between the expected true score curves on Testlet 1 could be detected. A summary of the differences observed is shown in Figure 6.5.

From this summary, we can see that since most of the focal group is concentrated between -1.5 and 0.5 on the proficiency dimension, the DIF in Testlet 4 has the most profound negative effect on African-Americans. Note that this negative effect is more than compensated for by the summed positive effects of the other testlets. This happy outcome is obscured by the apparent large negative

Table 6.6. *LSAT reading comprehension summary of White–African-American DIF*

Model	−2loglikelihood	# of free parameters
NO DIF	8,224	33
3 & 4 DIF	7,528	50
1, 3, & 4 DIF	7,438	57
2, 3, & 4 DIF	7,422	58
ALL DIF	7,408	64

DIF observable at high levels of proficiency. But because of the location of African-American proficiency distribution, this negative effect is substantially ameliorated.

Table 6.6 is a summary of the fit statistics that characterized these analyses. This table also includes the 1, 3, & 4 DIF analysis, which was not helpful.

This concludes the DIF analysis for one focal group. The same procedure was followed for four other focal groups (Hispanics, Asian-Americans, Women, and Canadians). In each case, we did exactly the same thing to arrive at a best fitting model. Of course, for each analysis we arrived at different results. For African-Americans, Testlets 3 and 4 were the passages with the most DIF. For other focal groups, other passages stood out. Is there some way to decide the worth of each passage over all of the focal groups? Given the spirit of DIF, looking at performance within a particular group, is such a question sensible?

On the off-chance that such a formulation is required, we summarized the total DIF for each focal group by summing the area between the expected true score curves, weighted by the proficiency distribution of the focal group (Wainer, 1993). Because this methodology is both obvious and unique, it is worthwhile taking a short diversion to explain it.

6.5.5. Standardized total impact

Suppose, in addition to detecting and representing the DIF in a testlet, we also wish to measure its amount. The following is a development of a model-based standardized index of impact that grows naturally out of the IRT approach taken in this study. The basic notion supposes that the impact of testlet j is the difference between the expected true score on that testlet if one were a member of the focal group versus what would have been the score if one was a member of the reference group (i.e., suppose one went to sleep *focal* and awoke *reference*; how much does that change one's expected true score?).

To answer this, we need a little notation:

Define:

$E_F(x_j|\theta)$ be the expected true score on testlet j for a member of the focal group, conditioned on having proficiency θ; and

$E_R(x_j|\theta)$be the expected true score on testlet j for a member of the reference group with the same proficiency.

Furthermore, let us define proficiency distributions for each group as $G_F(\theta)$ and $G_R(\theta)$,respectively. Lastly, suppose there are N_F and N_R individuals in each group. Now we can get on with the derivation of the indices of standardized impact.

When there is DIF and a person changes group membership overnight, their expected true score on the testlet changes. Thus, the amount of impact is

$$E_F(x_j|\theta) - E_R(x_j|\theta).$$

But, this must be weighted by the distribution of all of those affected, specifically, by $G_F(\theta)$. Thus, we will define the standardized index of impact, $T(1)$ as

$$T(1) = \int_{-\infty}^{\infty} [E_F(x_j|\theta) - E_R(x_j|\theta)]\mathrm{d}\,G_F(\theta).$$

Obviously, this is the average impact for each person in the focal group. Note that this is bounded regardless of the model used to generate the expected values. This index comes close to characterizing what we want. Indeed, for many purposes, this may be just right.

Is this always the right thing to be looking at? The amount of a testlet's impact depends on the choice of the focal group. An item might be unjust to one focal group, but just fine for another. Thus, it might be important for purposes of comparison to have a measure of *total impact*, or $T(2)= N_F T(1)$, or

$$T(2) = N_F \int_{-\infty}^{\infty} [E_F(x_j|\theta) - E_R(x_j|\theta)]\mathrm{d}\,G_F(\theta).$$

The concept of *total impact* is often a useful one in test construction. Consider that one constraint in test construction might be to choose a subset of testlets from a pool such that the total impact is minimized. Dorans (personal

Table 6.7. *Mean standardized impact*

Testlet	Focal group					Unweighted means	Focal weighted means
	Females	African-Americans	Hispanics	Asians	Canadians		
1	−0.3	−0.3	0.0	−0.1	−0.1	−0.2	−0.24
2	−0.1	−0.1	0.0	−0.2	−0.2	−0.1	−0.11
3	0.1	0.4	0.3	−0.2	−0.3	0.2	0.09
4	0.4	0.5	0.3	0.2	0.3	0.3	0.38
			Weights (N's)				
Focal	18,716	2,971	2,103	2,633	2,294		
Reference	22,022	32,778	32,778	32,778	41,302		
Total	40,738	35,749	34,881	35,411	43,596		

communication, October 26, 1991) describes a "melting pot" reference population, which is made up of all of the various focal groups. He suggested that one might then calculate the total impact (using an index such as $T(2)$) for each group relative to the whole. The operational testlet pool might be the one of requisite size that minimizes total impact (summed over the entire examinee population).

The concept of total impact allows us to consider the situation in which one testlet has only a small amount of sex *DIF* but no Aztec *DIF*, whereas another testlet might have more Aztec *DIF* but no sex *DIF*. This method allows us to choose between them on the basis of their total effect (sex groups are typically much larger than the number of LSAT taking Aztecs).

Table 6.7 shows the standardized impact for each testlet for each focal group. Testlet 1 is the one that seems to disadvantage these groups most uniformly, whereas Testlet 4 is the most disadvantageous for the reference groups. The figures provided under each focal group are what we have defined as $T(1)$. These are summed across focal groups in the column labeled "Unweighted means." The last column "Focal weighted means" is the sum of the $T(2)$ statistics; the weights used are at the bottom of the table. From this column, we see a more precise measure of the effect of testlet DIF. It averages about one fourth of a point against the various focal groups on Testlet 1 and about four tenths of a point in favor of the various focal groups on Testlet 4.

This completes the analysis for the reading comprehension section of these forms of the LSAT when comparing White and African-American examinees. For the purposes of all of these analyses, the two forms examined were merged by ignoring order effects and joining data from identical testlets.

6.6. Conclusions

This analysis has shown that the testlet structure of the reading comprehension and analytic reasoning sections of the LSAT has a significant effect on the statistical characteristics of the test. The testlet-based reliability of these two sections obtained in these analyses is considerably lower than what was previously calculated under the inaccurate assumption of local conditional independence. We believe that this discovery should change current practice. At a minimum, any reporting of section reliability ought to be modified to reflect our current knowledge. Moreover, analyses of current and future forms of these sections of the LSAT ought to model explicitly the testlet structure before calculating section reliability. Because of the test's overall length, we don't believe that the total test reliability that is reported is far enough wrong to have serious need for correction.

A second possibility would be to boost the section reliabilities up to the levels that were previously thought to hold through one of two modifications: (a) increase the length of the two testlet sections (six passages in reading comprehension and eight scenarios in analytic reasoning), or (b) use such statistical procedures as empirical Bayes estimation to "borrow" reliability for each section from the other sections (Thissen & Wainer, 2001, Chapter 9). Choice (a) would cause a substantial increase in testing time; choice (b) is practically free. Our experience with the use of procedures of this sort leads us to believe that the use of other sections to bolster the section scores would be substantial enough to yield section scores of sufficient statistical stability to justify reporting them separately. This may be a welcome addition to the LSAT for both law schools and examinees. Since the marginal cost of doing this is essentially nothing, we believe that it is an idea worth serious and immediate consideration. Another approach, a fully Bayesian method, has some distinct advantages and will be discussed in Part II.

While the balance of this book discusses fully Bayesian methods where parameters of prior distributions (e.g., the mean and variance of the distribution of the difficulty parameters) are assumed to be unknown and have their own prior distributions, there is an active area of research in which these parameters are estimated from the data and treated as fixed and known. These set of methods are known as empirical Bayes (Efron & Morris, 1975) methods.

The term empirical Bayes stems from the fact that one is still being Bayesian, that is, putting priors on the parameters of the baseline model (in our case the parameters of the IRT model): however, one is "empirically" estimating their values from the data. To be more technical, this is what Morris (1983) would describe as parametric empirical Bayes (assuming a parametric prior) in contrast to "empirical Bayes" methods that nonparametrically estimate the prior.

To estimate parameters in an empirical Bayes way, this is accomplished, typically, using maximum marginal likelihood (MML, Bock & Aitkin, 1981) procedures in which one first integrates over the distribution of the parameter, yielding a marginal distribution conditional on the prior parameter values, and then maximizes that function.

Empirical Bayes methods, however, do have a limitation that have reduced their utility, to some extent, in comparison to the fully Bayesian methods described here. In particular, by estimating the prior parameter values and treating them as fixed and known, there is a risk of underestimating the true variability in the system. For those parameter, estimates have uncertainty that fully Bayesian methods take into account, but is missed in empirical Bayes methods. However, empirical Bayes methods have faster computational properties that are sometimes of practical importance.

On a practical note, empirical Bayes methods can be run by the computer program SCORIGHT (described later in this book). One aspect of this software is its capacity to hold a set of parameters of the model at some fixed values. If one were to fix those parameters at the values estimated from an MML procedure, the fully Bayes SCORIGHT software would yield empirical Bayes results.

The DIF analyses provide reassurance. The size of the differences in the expected true score curves that was detected as statistically significant was very small. This ensures us of the impressive statistical power of this methodology. The size of these differences in the worst case was small enough to suggest that no serious problems of fairness exist. Moreover, even these small differences were reduced to almost nothing through the balancing across testlets. This balance cannot be swallowed whole. Because performance on the test section itself determined the stratifying variable, the overall balance (zero overall DIF) is almost tautological. That the balancing works as well as it does at all levels of examinee proficiency is not mathematically determined. It is one sign of a fair test.

Is this all? No. It is disquieting to note that the reading passage with the greatest impact on all focal groups was one where general topic was constitutional law. If ever there was a population of examinees for whom this topic is a fair one to include on a test it is the LSAT population. Yet, it shows up as the worst one. Why? One explanation is that it is a perfectly fair item. Perhaps, it is the other three passages that have DIF in favor of the various focal groups. Since we are using an internal criterion to measure DIF (to stratify examinee performance), such a conclusion would yield the observed results.

The only way to determine which of these two hypothetical interpretations is more nearly correct is through the use of more information. The most obvious source of information would be validity studies. One prospective study would be to see how predictive of success in law school is each of the passages. One can analyze such data in a way faithful to the structure of the test is by using law school grades as the stratifying variable in a DIF study. Then see which of the passages shows DIF and in which direction. When a validity criterion is used as the stratifying variable in a DIF study, we are no longer studying DIF; we are studying bias (in both the statistical and pejorative sense).

What we observe is not nature itself but nature exposed
to our method of questioning.
Werner Heisenberg (1958)

This investigation, based as it was on the testlet structure of half of the LSAT and polytomous IRT models, provides a glimpse into what is achievable with these psychometric tools. In Heisenberg's sense, we are using a more delicate and more rigorous method to question our data. Using such a tool has given us a deeper understanding of the structure of the test and its measurement characteristics. In addition to the "bad news" that two of the test's sections appear to be considerably less reliable than was previously thought, we found that through a more efficacious weighting of items, the logical reasoning sections were *more* reliable. The use of these sorts of modern measurement models can allow us to extract as much information as there is within each examinee's responses. Of at least equal importance, they also provide a more accurate estimate of the error that remains. These procedures have not yet become widely used operationally, and so despite their theoretical advantages, we do not have the same enormous experience with them as has accumulated with traditional procedures. However, with added experience, testlet-based IRT procedures show great promise.

This promise is limited, however, as the approach we described here is suboptimal in at least two respects. First, it relies on the sum score within each testlet. Thus, two examinees who get a score of 3 are considered indistinguishable regardless of which ones they got correct. If there is information in the pattern of responses, this method loses it. Second, each testlet is considered as a unit and if it is changed by, for example, changing one item that lies within it, it must be completely recalibrated. This means that modern adaptive algorithms that construct tests on the fly, one item at a time, cannot be used. In Chapter 7, we describe a new approach that provides a solution to these two problems and yields the beginning of a general psychometrics for testlets.

Questions

1. What sorts of errors emerge if we fit a test whose items are not conditionally independent with a scoring model that assumes it?
2. What is DIF?
3. What advantages are conferred in using a testlet approach to DIF?
4. What is the likelihood ratio test, and how can it be used to test for DIF?
5. How can the concept of standardized total impact be used in an operational testing program?

References

American Psychological Association. (1966). *Standards for educational and psychological tests and manuals*. Washington, DC: Author.

American Psychological Association. (1974). *Standards for educational and psychological tests*. Washington, DC: Author.

American Psychological Association. (1985). *Standards for educational and psychological testing*. Washington, DC: Author.

Anastasi, A. (1961). *Psychological testing* (2nd ed.). New York: Macmillan.

Birnbaum, A. (1968). Some latent trait models and their use in inferring an examinee's ability. In F. M. Lord & M. R. Novick (Eds.), *Statistical theories of mental test scores* (pp. 392–479). Reading, MA: Addison-Wesley.

Bock, R. D. (1972). Estimating item parameters and latent ability when responses are scored in two or more latent categories. *Psychometrika, 37*, 29–51.

Bock, R. D., & Aitkin, M. (1981). Marginal maximum likelihood estimation of item parameters: An application of an EM algorithm. *Psychometrika, 46*, 443–459.

Brown, W., & Thomson, G. H. (1925). *The essentials of mental measurement* (3rd ed.). London: Cambridge University Press.

Chang, H.-H., & Mazzeo, J. (1994). The unique correspondence of item response functions and item category response function in polytomously scored item response models. *Psychometrika, 59*, 391–404.

Crumm, W. L. (1923). Note on the reliability of a test, with a special reference to the examinations set by the College Entrance Board. *The American Mathematical Monthly, 30*(6), 296–301.

Dorans, N. J., & Holland, P. W. (1993). DIF detection and description: Mantel-Haenszel and standardization." In P. W. Holland & H. Wainer (Eds.), *Differential item functioning* (chap. 3, pp. 35–66). Hillsdale, NJ: Erlbaum.

Efron, B., & Morris, C. (1975). Data analysis using Stein's estimator and its generalizations, *Journal of the American Statistical Association, 70*, 311–319.

Green, B. F., Bock, R. D., Humphreys, L. G., Linn, R. L., & Reckase, M. D. (1984). Technical guidelines for assessing computerized adaptive tests. *Journal of Educational Measurement, 21*, 347–360.

Guilford, J. P. (1936). *Psychometric methods* (1st ed.). New York: McGraw-Hill.

Gulliksen, H. O. (1950). *Theory of mental tests.* New York: Wiley. (Reprinted, 1987, Hillsdale, NJ: Erlbaum)

Heisenberg, W. (1958). *Physics and philosophy*. New York: Harper & Row.

Holland, P. W. (1990). On the sampling theory foundations of item response theory models. *Psychometrika, 55*, 577–601.

Holland, P. W., & Thayer, D. T. (1988). Differential item performance and the Mantel–Haenszel procedure. In H. Wainer & H. Braun (Eds.), *Test validity* (pp. 129–145). Hillsdale, NJ: Erlbaum.

Holland, P. W., & Wainer, H. (Eds.). (1993). *Differential item functioning*. Hillsdale, NJ: Erlbaum.

Holzinger, K. (1923). Note on the use of Spearman's prophecy formula for reliability. *The Journal of Educational Psychology, 14*, 302–305.

Humphreys, L. G. (1962). The organization of human abilities. *American Psychologist, 17*, 475–483.

Humphreys, L. G. (1970). A skeptical look at the factor pure test. In C. E. Lunneborg (Ed.), *Current problems and techniques in multivariate psychology: Proceedings of a conference honoring Professor Paul Horst* (pp. 23–32). Seattle: University of Washington.

Humphreys, L. G. (1981). The primary mental ability. In M. P. Friedman, J. P. Das, & N. O'Connor (Eds.), *Intelligence and learning* (pp. 87–102). New York: Plenum Press.

Humphreys, L. G. (1986). An analysis and evaluation of test and item bias in the prediction context. *Journal of Applied Psychology, 71*, 327–333.

Kelley, T. L. (1924). Note on the reliability of a test: A reply to Dr. Crumm's criticism. *The Journal of Educational Psychology, 15*, 193–204.

Kendall, M. G., & Stuart, A. (1967). *The advanced theory of statistics* (Vol. II). London: Charles Griffin.

Lewis, C., & Sheehan, K. (1990). Using Bayesian decision theory to design a computerized mastery test. *Applied Psychological Measurement, 14*, 367–386.

Lord, F. M. (1980). *Applications of item response theory to practical testing problems*. Hillsdale, NJ: Erlbaum.

Lord, F. M., & Novick, M. (1968). *Statistical theories of mental test scores*. Reading, MA: Addison-Wesley.

Morris, C. N. (1983). Parametric empirical Bayes inference: Theory and applications with discussion. *Journal of the American Statistical Association, 78*, 47–65.

Neyman, J., & Pearson, E. S. (1928). On the use and interpretation of certain test criteria for purposes of statistical inference. *Biometrika, 20A*, 174–240, 263–294.

Roznowski, M. (1988). Review of *test validity. Journal of Educational Measurement, 25*, 357–361.

Samejima, F. (1969). *Estimation of latent ability using a response pattern of graded scores* (Psychometric Monograph Supplement 4, 17, pt. 2) Richmond, VA: Psychometric Society.

Shealy, R., & Stout, W. (1993). A model-based standardization approach that separates true bias/DIF from group ability differences and detects test bias/DIF as well as item bias/DIF. *Psychometrika, 58*, 159–194.

Sireci, S. G., Thissen, D., & Wainer, H. (1991). On the reliability of testlet-based tests. *Journal of Educational Measurement, 28*, 237–247.

Swaminathan, H., & Rogers, H. J. (1990). Detecting differential item functioning using logistic regression procedures. *Journal of Educational Measurement, 27*, 361–370.

Thissen, D. (1990). MULTILOG user's guide (Version 6.0). Mooresville, IN: Scientific Software.

Thissen, D. (1991). Multilog user's guide [Computer program]. Chicago, IL: Scientific Software.

Thissen, D. (1993). Repealing rules that no longer apply to psychological measurement. In N. Frederiksen, R. J. Mislevy, & I. Bejar (Eds.), *Test theory for a new generation of tests* (pp. 79–97). Hillsdale, NJ: Erlbaum.

Thissen, D., & Steinberg, L. (1984). A response model for multiple choice items, *Psychometrika, 49*, 501–519.

Thissen, D., & Steinberg, L. (1988). Data analysis using item response theory. *Psychological Bulletin, 104*, 385–395.

Thissen, D., Steinberg, L., & Fitzpatrick, A. R. (1989). Multiple-choice models: The distractors are also part of the item. *Journal of Educational Measurement, 26*, 161–176.

Thissen, D., Steinberg, L., & Mooney, J. (1989). Trace lines for testlets: A use of multiple-categorical response models. *Journal of Educational Measurement, 26*, 247–260.

Thissen, D., Steinberg, L., & Wainer, H. (1988). Use of item response theory in the study of group differences in trace lines. In H. Wainer & H. Braun (Eds.), *Test validity* (pp. 147–169). Hillsdale, NJ: Erlbaum.

Thissen, D., Steinberg, L., & Wainer, H. (1993). Detection of differential item functioning using the parameters of item response models. In P. W. Holland & H. Wainer (Eds.), *Differential item functioning* (pp. 67–113). Hillsdale, NJ: Lawrence Erlbaum Associates.

Thissen, D., & Wainer, H. (Eds.). (2001). *Test scoring*. Hillsdale, NJ: Erlbaum.

Thorndike, R. L. (1951). Reliability. In E. F. Lindquist (Ed.), *Educational measurement. Washington*, DC: American Council on Education.

Traub, R. E., & Rowley, G. L. (1991). Understanding reliability. *Educational Measurement: Issues and Practice, 10*, 37–45.

Wainer, H. (1993). Model-based standardized measurement of an item's differential impact. In P. W. Holland & H. Wainer (Eds.), *Differential item functioning* (pp. 121–135). Hillsdale, NJ: Erlbaum.

Wainer, H., & Kiely, G. L. (1987). Item clusters and computerized adaptive testing: A case for testlets. *Journal of Educational Measurement, 24*, 185–202.

Wainer, H., & Lewis, C. (1990). Toward a psychometrics for testlets. *Journal of Educational Measurement, 27*, 1–14.

Wainer, H., Sireci, S. G., & Thissen, D. (1991). Differential testlet functioning: Definitions and detection. *Journal of Educational Measurement, 28*, 197–219.

Wang, M. D., & Stanley, J. C. (1970). Differential weighting: A review of methods and empirical studies. *Review of Educational Research, 40*, 663–705.

Yen, W. M. (1992, April). *Scaling performance assessments: Strategies for managing local item dependence*. Invited address presented at the annual meeting of the National Council on Measurement in Education, San Francisco, CA.

Zwick, R. (1990). When do item response function and Mantel–Haenszel definitions of differential item functioning coincide? *Journal of Educational Statistics, 15*, 185–198.

PART II

Bayesian testlet response theory

Recapitulation and introduction

The invention of short multiple-choice test items provided an enormous technical and practical advantage for test developers; certainly, the items could be scored easily, but that was just one of the reasons for their popular adoption in the early part of the 20th century. A more important reason was the potential for an increase in validity because of the speed with which such items could be answered. This meant that a broad range of content specifications could be addressed, and hence an examinee no longer needed to be heavily penalized because of the unfortunate choice of a question requiring a longer-to-answer constructed response. These advantages, as well as many others (see Anastasi, 1976, pp. 415–417), led the multiple-choice format to become, by far, the dominant form used in large-scale standardized mental testing for the past century. Nevertheless, this breakthrough in test construction, dominant at least since the days of Army *alpha* (1917), is currently being reconsidered.

Critics of tests that are made up of large numbers of short questions suggest that de-contextualized items yield a task that is abstracted too far from the domain of inference for many potential uses, and hence would have low predictive validity for most practical issues. For several reasons, only one of them as a response to this criticism, variations in test theory were considered that would allow the retention of the short-answer format while eliminating the shortcomings expressed by those critics. One of these variations was the development of item response theory (IRT), an analytic breakthrough in test scoring. A key feature of IRT is that examinee responses are conceived of as reflecting evidence of a particular location on a single, underlying latent trait. Thus, the inferences that are made are about that trait and not constrained to what some might think to be atomistic responses to specific individual items. A second approach, with a long and honored history, was the coagulation of items into coherent groups that could be scored as a whole. As defined earlier, such groups of items are often referred to as testlets (Wainer & Kiely, 1987), since they are typically longer than a single item (providing greater context) yet shorter than an entire test. These introductory remarks provide a history and motivation for our approach to the inclusion of testlets into a coherent psychometric model for test calibration and scoring.

Previously, we described three reasons for using testlets; there are others. One reason is to reduce concerns about the atomistic nature of single independent small items. A second is to reduce the effects of context[1] in adaptive testing; this is accomplished because the testlet, in a very real sense, carries its context with it (except for the items at the boundaries). The third reason is to improve the efficiency (information per unit time) of testing when there is an extended stimulus (e.g., a reading passage, a diagram, or a table) associated with an item. A substantial part of an examinee's time is spent processing the information contained in the stimulus material, thus greater efficiency is achieved if more than a single bit of information is collected from each stimulus. Obviously, diminishing returns eventually sets in; one cannot ask 200 independent questions about a 250-word passage. Usually, the depths of meaning carried by such a passage are considered plumbed by four to six questions. More recent trends have yielded longer passages, often with as many as 10 to 12 items per passage.

Traditionally, the items that made up testlets were scored as if they were independent units just like all other items. Research has shown that if the excess within-testlet variation is irrelevant to the measurement of the construct of interest, scoring that ignores local dependence, tends to overestimate the precision of measurement obtained from the testlet (Sireci, Wainer, & Thissen, 1991; Wainer & Thissen, 1996; Yen, 1993). Happily, with testlets of modest length (e.g., a four-item testlet and a 250-word passage), the effect may be minimal; albeit this phenomenon is not necessarily correlated with the number of questions per testlet. As described and illustrated in Chapter 6, a scoring method that accounts for the item information more accurately (than conditionally independent atomistic responses) treats the testlet as a single item and scores it polytomously. This approach has worked well (Wainer, 1995) in a broad array

[1] We use the term "context" here very narrowly. For our purposes, we define the *context of an item* by the items that are immediately around it. Usually, we mean what item immediately precedes and follows it. Thus, later when we say that a testlet carries its own context, we mean that except for the first and last items in the testlet, each item of the testlet always has the same context.

of situations and is a good practical approach. But there are two circumstances where something more is needed.

Circumstance 1: If you need to extract more information from the testlet. When a polytomous IRT model is used to fit a testlet, the testlet score is represented by the number correct (it could also be a weighted-number correct). As evidenced by the number of successful applications of the Rasch model, such a representation is often good enough. Yet, there is some information in the exact pattern of correct scores. To extract this information, a more complex model is required.

Circumstance 2: Ad hoc testlet construction within a computerized adaptive test (CAT). A common current practice is to build testlets on the fly within the context of a CAT. For example, in the CAT's item pool, there might be an item stimulus (e.g., reading passage) and 15 associated items. There might never have been any intent that all 15 items would be presented to any specific examinee; indeed, quite often those items were written with the express notion that some of the items are incompatible with one another (i.e., one might give the answer to another, or one might be an easier version of another). The item-selection algorithm chooses the testlet stimulus (e.g., the passage) and then picks one item on the basis of item content, the psychometric properties of that item, and the previous examinee responses. After the examinee responds to that item, the examinee's proficiency estimate is updated and a second item from those associated with that stimulus is chosen. This process is continued until a prespecified number of items have been presented and the item-selection algorithm moves on. At the present time, the algorithm behaves as if all of the items it chooses within the testlet are conditionally independent. As with any algorithm, this form of testlet-based item selection creates missing data, that is, the unobserved responses to those testlet items not administered. We discuss this issue in more detail in Chapter 12.

These two testing circumstances have been occurring more frequently of late because of the increasing popularity of adaptive testing and the strong desire for test construction units (the testlets) to be substantial and coherent. If test formats continue into the future as they have existed in the past, with testlets being of modest length (4–6 items per testlet), we do not expect that the unmodeled local dependence would prove to be a serious obstacle in the path of efficient testing or accurate scoring; yet examples exist such as Wang, Bradlow, and Wainer (2002), where ignoring testlet dependence even in moderate size testlets makes a difference. But as testlets grow longer, explicit modeling of their effects would be needed. The approach we describe next is a parametric framework that encompasses many alternative test formats likely to be seen in practice.

In short, it is a psychometric theory of testlets – what we term testlet response theory (TRT).

This part develops TRT in a stepwise fashion, moving from a simple form of test model to increasingly complex ones. The goal is to provide a test scoring model for increasingly general test formats. We begin with a model for a test made up of testlets composed of binary items in which the probability of guessing is not high enough to worry about; the testlet analog to the 2-PL model. We then generalize the situation to allow for guessing: the analog to the 3-PL model. Next, we derive a model that allows for items that are scored polytomously. At all stages, we try to remain completely general in that we allow testlets to be made up of any number of items, including, in the degenerate case, only a single item. In this latter situation, each of the testlet models specializes to the corresponding standard IRT model. In addition, testlets need not be homogeneous with respect to item type; a testlet might contain a mixture of 2-PL, 3-PL, and polytomously scored items. Our goal was to let the test developer(s) decide which test structure would best measure the constructs of interest, and have the test scoring model be flexible enough to handle that structure properly.

Finally, we embed within every testlet model the possibility of including covariate information. A discussion of how this is done and what information might be gained from the inclusion of covariates is postponed until after we discuss, in turn, each of the test scoring models, but this capability could be used with all of the models.

More specifically, the rest of this part is constructed as follows:

Chapter 7 provides an overview of the general psychometric approach including both some of the modeling structures and an outline of the estimation methodology. Greater details of both will be given as they are needed.

Chapter 8 implements and illustrates this approach with the testlet version of the well-known two-parameter logistic IRT model (2-PL). In this chapter, we provide an in-depth description of the probability model into which the psychometric model is embedded as well as details of the estimation method.

Chapter 9 generalizes the model to the three-parameter logistic IRT model (3-PL), which allows for the possibility of an examinee getting an item correct accidently.

Chapter 10 expands the methodology still further by generalizing it to also include testlets that are made up of polytomously scored items.

Chapter 11 moves TRT ahead by showing how the model could include covariates. A typical IRT analysis answers such questions as "How difficult was this item?" and "How able was this examinee?" Through the inclusion of covariate information, it could allow the analysis to shed light on "why?" "Why was this item so difficult?" and "Why did this person score so well?" We believe that this capability represents, in prospect, a remarkably useful advance in psychometric theory.

Chapter 12 opens the can of worms of the real-life situation in which the data sets are incomplete. It explores alternative approaches to dealing with missing data of various sorts and their consequences.

References

Anastasi, A. (1976). *Psychological testing* (4th ed.). New York: Macmillan.

Sireci, S. G., Wainer, H., & Thissen, D. (1991). On the reliability of testlet-based tests. *Journal of Educational Measurement, 28*, 237–247.

Wainer, H. (1995). Precision & differential item functioning on a testlet-based test: The 1991 Law School Admissions Test as an example. *Applied Measurement in Education, 8*(2), 157–187.

Wainer, H., & Kiely, G. (1987). Item clusters and computerized adaptive testing: A case for testlets. *Journal of Educational Measurement, 24*, 185–202.

Wainer, H., & Thissen, D. (1996). How is reliability related to the quality of test scores? What is the effect of local dependence on reliability? *Educational Measurement: Issues and Practice, 15*(1), 22–29.

Wang, X., Bradlow, E. T., & Wainer, H. (2002). A general Bayesian model for testlets: Theory and applications," *Applied Psychological Measurement*, *26*(1), 190–128. Also listed as ETS GRE Technical Report 98-01.

Yen, W. (1993). Scaling performance assessments: Strategies for managing local item dependence. *Journal of Educational Measurement, 30*, 187–213.

7

A brief history and the basic ideas of modern testlet response theory

7.1. Introduction. Testlets as a multidimensional structure

One useful way to think about a test built of k testlets utilizes factor analytic vocabulary (Gessaroli & Folske, 2002). First, we note that all of the testlets contain information about the underlying trait of interest – a general factor. However, each of the items within a testlet also has some connection with the other items in that testlet, over and above the general factor – a testlet-specific factor. Thus, the factor structure of such a test has $k + 1$ orthogonal factors; a general factor, plus one factor for each testlet.

The story that describes a theoretical test structure, and its associated mode of analysis based on this concept, begins in 1937 with the publication of Karl Holzinger and Francis Swineford's bi-factor method. Holzinger and Swineford (1937) conceived of the bi-factor method as an extension of Spearman's two-factor model, and its character matched the conceptual structure that Thurstone had for his primary mental abilities (Thurstone, 1938). Specifically, they thought of the underlying structure of a bank of tests as consisting of a single general factor and each set of tests loading on its own specific factor. They illustrated this with a bank of 15 tests that all load on a single general factor of ability, but, in addition, had four tests that loaded on a spatial factor, three other tests that loaded on a mental speed factor, four tests on a motor speed factor, and the remaining four tests on a verbal factor. The bi-factor solution constructed all of the factors to be orthogonal, and was so named because each test loaded on exactly two factors – a general one and a specific one related to a subset of the tests that define that specific factor.

Their goal in proposing this method was three-fold. First, they wanted an analytic structure that would allow all of the information in the 15 tests that bore on the general factor to be extracted. Second, they wanted simultaneously to separate that aspect of the residual variance that was in common among all of the tests on the same subject. Third, they wanted to simplify calculations, for the more general factor model was very difficult to compute at that time. In formulating the bi-factor model, they accomplished all three goals. In fact, with respect to the third goal, they concluded their paper with the comment "The bi-factor analysis illustrated above is not only very simple, but the calculation is relatively easy as compared with other methods. The total time for computation, done by one person, was less than ten hours for the present example" (p. 54).[2]

It is straightforward to translate their work into the terminology we have used throughout this book; we might say that the test bank could be thought of as a single test composed of 15 testlets. Holzinger and Swineford (1937) were thinking of the subtest as the fungible unit of the whole test, and all calculations were done with the subtest score. Once the subtest scores were obtained, the items that made up the subtests were largely forgotten.[3] In this way, this is directly analogous to the approach discussed in Chapter 6, in which each testlet was treated as a polytomously scored item. And, just as the polytomous scoring of a testlet-based test has provided a starting point for testlet response theory (TRT), so too have Holzinger and Swineford (1937) provided a starting point for a more general approach to TRT. But to move beyond this to a testlet structure in which the individual items are an integral part of the analysis requires that we jump forty-five years forward in time to a development proposed by Gibbons and Hedeker (1992).

7.2. Advances in estimation and computing: essential developments to allow us to include items as part of the testlets

There was a substantial amount of work done in factor analysis in the forty-five years between Holzinger and Swineford (1937) and Gibbons and Hedeker (1992); among the most influential sources are Harmon (1975), Jörsekog (1974); Jöreskog and Sörbom (1979), and McDonald (1985, especially Chapter 3). Gibbons and Hedeker (1992) provided an especially clear generalization of the

[2] By that standard, the testlet analyses that are subsequently, described done by the computer program SCORIGHT, are easier still. There are none that we have done, even with extended burn-in, that took longer than 10 hours. We might henceforth dub the one person, 10-hour time as a single unit – the HS – in honor of Holzinger and Swineford. If an analysis takes 20 hours, we shall indicate it as "2HS". If it takes 5 hours, it is a "0.5HS". Currently, most analyses are in the centi-HS range (HS/100), but with faster processors and improved computational algorithms, we anticipate milli-HS analyses in the future.

[3] We assume that they took this approach because the increase in computational burden doing this analysis at the item level would have made their approach impractical with the computing capabilities available at that time.

Holzinger and Swineford approach to allow the analysis to take place at the item level. To accomplish this, they were able to lean on a number of important prior developments. The groundwork was laid by Ledyard Tucker (1958) in his development of interbattery factor analysis that generalized the bi-factor approach to allow the inclusion of individual items. Jöreskog (1969) demonstrated the practicality of this approach within a framework of confirmatory factor analysis in which the parameters were estimated by maximum likelihood. But Gibbons and Hedeker's development was computationally simpler and faster because they were able to use Bock and Aitken's (1981) estimation methodology that maximizes the marginal likelihood, which was itself made possible through the EM algorithm (Dempster, Laird, & Rubin, 1977) as well as the technology of full information factor analysis (Bock, Gibbons, & Muraki, 1988).

The normal ogive model proposed by Gibbons and Hedeker (1992) is formally very close to the approach we espouse here. Before we elaborate, it is worthwhile to note that one of the most important characteristics of the almost seventy-year-long pathway from Holzinger and Swineford through Gibbons and Hedeker to our presentation here is the method of estimation. Working with paper and pencil or rudimentary desktop calculators meant that all sorts of shortcuts must be taken to make any sort of latent trait analysis tractable. By the early 1990s, high-speed computing and the EM algorithm were already ubiquitous, and so shortcuts were less important. The fact that the bi-factor model could be extended easily to include items made Gibbons and Hedeker's ingenious development inevitable.

But a fresh Bayesian wind was blowing at about the same time that Gibbons and Hedeker wrote their paper. It has long been understood that there were important advantages to be obtained for making inferences from data if we could use probability models for both the quantities we observe and the quantities about which we wish to learn. What was lacking were the practical means for doing so. These practical methods, described in Gelman et al. (1995), rest on a foundation of Monte Carlo estimation. The initial work on estimation using Monte Carlo procedures was described more than forty years earlier (Metropolis & Ulan, 1947), but lack of computing power prevented its widespread use. Metropolis et al. (1953) demonstrated its potential, even with the primitive computing equipment then available, but it had too little practical power without big computing. Twenty years was to elapse before Hastings (1970) polished up the Metropolis algorithm to yield the eponymous method in broad use today. Almost two decades more went by until a rush of methods and applications appeared. Tanner and Wong (1987) showed us how to use data augmentation, in combination with the EM algorithm, to compute posterior distributions. Gelfand and Smith (1990) expanded this in the calculation of marginal densities.

Then James Albert (Albert, 1992; Albert & Chib, 1993) brought the technology closer to IRT with his papers on the use of this new technology in estimating normal ogive response functions within a fully Bayesian framework, setting the stage for our use of it in the development of a general testlet response theory (Bradlow, Wainer, & Wang, 1999).

7.3. Why Bayesian?

The embedding of a test scoring model, whether based on items (IRT) or testlets (TRT), within a full probability model in which a joint probability distribution for all observable and unobservable quantities yields many advantages. The most obvious one is that it provides us with a mechanism to contain, within the model, a representation of our knowledge of the underlying structure of the test as well as the way that the data were collected. There is no easier way of formally including everything we know into the scientific mechanism from which we would be drawing inferences. A Bayesian framework yields many other benefits as well, not all of them known yet, but with the development of Monte Carlo estimation methods such as the Gibbs sampler,[4] these methods have recently become practical. Using a Bayesian approach requires care and computer time, but not genius. This was well phrased by Donald Rubin (personal communication, August 8, 2003):

> If you are super clever, perhaps you can get an answer that has good frequentist properties to hard problems by thinking really hard, whereas just by being relatively careful you can do so by being Bayesian. An analogy I am fond of is to consider how middle school children might solve mathematical word problems: the frequentist approach is analogous to trying to figure them out by being smart, the Bayesian approach is analogous to having learned algebra, plugging in and solving. When the problem is easy, if you are smart enough you don't need the algebra, but eventually algebra helps, no matter how smart you are.

What are the difficult problems that a Bayesian approach allows us to solve easily? A sampling of answers to this is contained in Part III of this book, but for now let us get a sniff of what is to come. A natural consequence of the Bayesian approach is that we do not just obtain a point estimate of each of the parameters of interest; instead, we get their entire posterior distribution. Thus possibly, complex practical problems are solved easily. For example, suppose you take a test with a specified passing score and you obtain a score somewhat below that passing score. A natural question would then be, "if I were

[4] In a discussion of image processing, Geman and Geman (1984) named this new approach as the Gibbs sampling.

to take the test again, without learning anything new, how likely is it, due just to the vagaries of testing, that I will pass?" With a Bayesian approach, we have the entire posterior distribution of your score, so to answer the question all we need do is sample from that posterior and count up the proportion of times that the sampled value is greater than the cut score. And, for greater precision in answering that question, one merely has to increase the size of the sample.

Contrast this solution with what we would do within the confines of traditional true score theory. First, we would calculate the reliability, and, remembering that the reliability ($\rho_{xx'}$) is equal to 1 minus the ratio of the error variance to the observed score variance, or

$$\rho_{xx'} = 1 - \left(\frac{\sigma_e^2}{\sigma_x^2}\right), \tag{7.1}$$

which allows us to calculate the standard error of measurement, σ_e, by solving Equation (7.1), thus getting

$$\sigma_e = \sigma_x(1 - \rho_{xx'})^{1/2}. \tag{7.2}$$

We can then combine the result in (7.2) with an untested assumption of normality to estimate the answer to the initial question by looking at the tail area under the normal cumulative density function (CDF). The accuracy of the answer depends on all of its component pieces; the normality assumption being the most problematic.

Compare the complexity of the two approaches; the latter requires that you understand normal curve theory, whereas the former only asks you to be able to count.

7.4. Modeling testlets: introducing the testlet parameter $\gamma_{id(j)}$

Moving from the estimation to the set of models that comprise this version of TRT requires two basic probability kernels that allow us to encompass both dichotomous and polytomous items. For dichotomous items, we utilize the 3-PL model:

$$P(Y_{ij} = 1) = c_j + (1 - c_j)\,\text{logit}^{-1}(t_{ij}), \tag{7.3}$$

and for polytomous items, we utilize the ordinal response model introduced by Samejima (1969):

$$P(Y_{ij} = r) = \Phi(d_r - t_{ij}) - \Phi(d_{r-1} - t_{ij}), \tag{7.4}$$

where Y_{ij} is the response of examinee i on item j;
c_j is the lower asymptote (guessing parameter) for dichotomous item j,
d_r and d_{r-1} are the latent cutoffs (score thresholds) for the polytomous items
logit $= \log(x/(1-x))$;
Φ is the normal CDF; and
t_{ij} is the latent linear predictor of score.

The two-parameter dichotomous items are a special case of the 3-PL model with $c_j = 0$. In this special case,

$$P(Y_{ij} = 1) = \text{logit}^{-1}(t_{ij}), \tag{7.5}$$

we model the extra dependence due to testlets by extending the linear score predictor t_{ij} from its standard form

$$t_{ij} = a_j(\theta_i - b_j), \tag{7.6}$$

where a_j, b_j, and θ_i have their standard interpretations as item slope, item difficulty, and examinee proficiency, to

$$t_{ij} = a_j(\theta_i - b_j - \gamma_{id(j)}), \tag{7.7}$$

with $\gamma_{id(j)}$ denoting the testlet effect (interaction) of item j with person i that is nested within testlet $d(j)$.

The extra dependence of items within the same testlet (for a given examinee) is modeled in this manner, as both items would share the effect $\gamma_{id(j)}$ in their score predictor. By definition, $\gamma_{id(j)} = 0$ for all independent items. Thus, to sum up the model extension here, it is the set of parameters $\gamma_{id(j)}$ that represents the difference between this model and standard approaches. It is around this model that we introduce a hierarchical Bayesian framework to combine all information across examinees, items, and testlets.

7.5. Summing up

The formal models for TRT that comprise the rest of this book are based on the three pieces described in this introduction:

1. The formal psychometric model that includes the extra parameter $\gamma_{id(j)}$ that captures the excess within-testlet covariation.
2. A full probability model that provides the joint probability distribution for all quantities of interest and is consistent with everything we know about how the test data were generated and collected.

3. A Markov chain Monte Carlo procedure that allows us to infer the posterior distribution (or summaries of it) of the parameters of the model.

All of these are contained in the computer program SCORIGHT (Wang, Bradlow, & Wainer, 2004), which is available from the authors free of charge.

In the chapters that follow, each of these three pieces will be elaborated upon and illustrated.

Questions

1. How can we characterize a testlet-based test in factor analytic terms?
2. What were the impediments to the adoption of a Bayesian approach to testing prior to 1990?
3. What are the advantages of a Bayesian approach to test theory?
4. How does the testlet model differ from standard IRT models?

References

Albert, J. H. (1992). Bayesian estimation of normal ogive item response curves using Gibbs sampling. *Journal of Educational Statistics, 17*, 251–269.

Albert, J. H., & Chib, S. (1993). Bayesian analysis of binary and polychotomous response data. *Journal of the American Statistical Association, 88*, 669–679.

Bock, R. D., & Aitkin, M. (1981). Marginal maximum likelihood estimation of item parameters: An application of an EM algorithm. *Psychometrika, 46*, 443–459.

Bock, R. D., Gibbons, R., & Muraki, E. (1988). Full information item factor analysis. *Applied Psychological Measurement, 12*, 261–280.

Bradlow, E. T., Wainer, H., & Wang, X. (1999). A Bayesian random effects model for testlets. *Psychometrika, 64*, 153–168.

Dempster, A. P., Laird, N. M., & Rubin, D. B. (1977). Maximum likelihood from incomplete data via the EM algorithm (with discussion). *Journal of the Royal Statistical Society, Series B, 39*, 1–38.

Gelfand, A. E., & Smith, A. F. M. (1990). Sampling-based approaches to calculating marginal densities. *Journal of the American Statistical Association, 85*, 398–409.

Gelman, A., Carlin, J. B., Stern, H. S., & Rubin, D. B. (1995). *Bayesian data analysis.* London: Chapman & Hall.

Geman, S., & Geman, D. (1984). Stochastic relaxation, Gibbs distributions, and the Bayesian restoration of images. *IEEE Transactions on Pattern Analysis and Machine Intelligence, 6*, 721–741.

Gessaroli, M. E., & Folske, J. C. (2002). Generalizing the reliability of tests comprised of testlets. *International Journal of Testingx, 2*, 277–295.

Gibbons, R. D., & Hedeker, D. R. (1992). Full information item bi-factor analysis. *Psychometrika, 57*, 423–436.

Harmon, H. H. (1976). *Modern factor analysis* (3rd ed). Chicago: University of Chicago Press.

Hastings, W. K. (1970). Monte Carlo sampling methods using Markov chains and their applications. *Biometrika, 54*, 93–108.

Holzinger, K. J., & Swineford, F. (1937). The bi-factor method. *Psychometrika, 2*, 41–54.

Jöreskog, K. G. (1969). A general approach to confirmatory maximum likelihood factor analysis. *Psychometrika, 34*, 183–202.

Jöreskog, K. J. (1974). Analyzing psychological data by structural analysis of covariance matrices. In D. H. Krantz, R. C. Atkinson, R. D. Luce, & P. Suppes (Eds.), *Contemporary developments in mathematical psychology* (Vol. II, pp. 1–56). San Francisco: W. F. Freeman.

Jöreskog, K. J., & Sörbom, D. (1979). *Advances in factor analysis and structural equation models*. Cambridge, MA: Abt Books.

McDonald, R. P. (1985). *Factor analysis and related methods.* Hillsdale, NJ: Erlbaum.

Metropolis, N., & Ulan, S. (1949). The Monte Carlo method. *Journal of the American Statistical Association, 44*, 335–341.

Metropolis, N., Rosenblith, A. W., Rosenblith, M. N., Teller, A. H., & Teller, E. (1953). Equation of state calculations by fast computing machines. *Journal of Chemical Physics, 21*, 1087–1092.

Samejima, F. (1969). *Estimation of latent ability using a response pattern of graded scores* (Psychometric Monograph Supplement 4, Part 2, No. 17). Richmond, VA: Psychometric Society.

Tanner, M. A., & Wong, W. H. (1987). The calculation of posterior distributions by data augmentation (with discussion). *Journal of the American Statistical Association, 82*, 528–550.

Thurstone, L. L. (1938). *Primary mental abilities* (Psychometric Monograph No. 1). Chicago: University of Chicago Press.

Tucker, L. R. (1958). An interbattery method of factor analysis. *Psychometrika, 23*(2), 111–136.

Wang, X., Bradlow, E. T., & Wainer, H. (2004). A user's guide for SCORIGHT (Version 3.0): A computer program for scoring tests built of testlets including a module for covariate analysis (ETS Technical Report RR-04-49). Princeton, NJ: Educational Testing Service.

8

The 2-PL Bayesian testlet model

8.1. Introduction

As described in the preceding chapters, the initial goal of testlet response theory (TRT) was to generalize existing item response theory (IRT) models so that in the case where no testlet dependence occurred, the model would specialize to the standard atomistic testing setting in which the assumption of conditional independence across items is commonly assumed and likely to hold. However, in those cases where multiple items are based upon a single stimulus (passage, table, etc. . .), and extra dependence across items does exist, the model is able to naturally account for this. A major advantage of this approach is that it allows us to use the familiar nomenclature of IRT, and thus aids in easy interpretation of the results, as users are familiar with the model, except for the addition of a "testlet term." In addition, standard and widely distributed programs such as MULTILOG 7 (Thissen Chen, & Bock, n.d.) could be used as both a validation check for our inferences as well as providing starting values for our procedures.

8.2. The model specification

In this vein, the initial TRT model developed in Bradlow, Wainer, and Wang (1999) had the following characteristics and serves as an important baseline model (which has subsequently been extended and described in Chapters 9–11). We first begin by describing the characteristics of the model, followed by computation under the model, a demonstration of the efficacy of the model, and finally extensions to this baseline model.

1. The model was designed to be fit to dichotomous test item data. As binary items are by far the most common form of item scores, this was an

important place to start; yet, as described in Part I, would be limiting if this were the only form of assessment available via TRT.

2. The model fit to the binary items is the 2-PL model, given by $P_{ij}(y_{ij} = 1|t_{ij}) = \text{logit}^{-1}(t_{ij}))$, where t_{ij} is the latent linear predictor of score, $y_{ij} = 0$ or 1, the response score from person i on item j, and $\text{logit}^{-1}(t_{ij}) = \exp(t_{ij})/(1 + \exp(t_{ij}))$. As the 2-PL model is a widely used model, it served well as the standard model.
3. The assumed model for t_{ij} was extended from the standard 2-PL model in which $t_{ij} = a_j(\theta_i - b_j)$ to one that incorporates testlet dependence – $t_{ij} = a_j(\theta_i - b_j - \gamma_{id(j)})$, where as before θ_i, a_j, and b_j are the person ability, item discrimination, and item difficulty, respectively. The additional term in the model, $\gamma_{id(j)}$, incorporates the extra dependence when two items j and j' are in the same testlet, that is, $d(j) = d(j')$.
4. The entire testlet model is embedded within a Bayesian framework that allows for sharing of information across persons, items, and testlets. In particular, the Bayesian hierarchical structure that we employ is
   i. $\theta_i \sim N(0, 1)$ to identify the model,
   ii. $\log(a_j) \sim N(\mu_a, \sigma_a^2)$,
   iii. $b_j \sim N(\mu_b, \sigma_b^2)$, and
   iv. $\gamma_{id(j)} \sim N(0, \sigma_\gamma^2)$

   with slightly informative normal-inverse gamma priors on the vector of means and variance components, respectively, to ensure proper posteriors.

We note that by utilizing this structure, two item responses for person i, j and $j'(j \neq j')$ that are both in testlet $d(j) = d(j')$, share the term $\gamma_{id(j)}$ in their latent linear predictor t_{ij}, and hence are more highly correlated under the model than items j and j' for which $d(j) \neq d(j')$ (as $\gamma_{id(j)}$ is assumed to be independent of $\gamma_{id(j')}$ for different testlets). Thus, the term $\gamma_{id(j)}$ is the one that yields the Bayesian TRT model, and when $\gamma_{id(j)} = 0$, we have the standard 2-PL model, albeit a Bayesian one. Of course, if we allowed $\sigma_a^2 \to \infty$, $\sigma_b^2 \to \infty$, and $\sigma_\gamma^2 \to \infty$, we would obtain the standard 2-PL model, and hence our desire to nest the standard practice is met.

The Bayesian aspect of the model, of special note, is that via the priors given above there is sharing of information across items where smaller values of σ_a^2 and σ_b^2 would cause greater shrinkage of a_j and b_j to the aggregate population values μ_a and μ_b, respectively; this holds similarly for person–testlet responses $\gamma_{id(j)}$ and σ_γ^2. This shrinkage is important as the amount of data for a given person–testlet combination is small, constrained as it is by the number of items in testlet $d(j)$. We next describe how computation is done under the model.

8.3. Computation under the Bayesian 2-PL TRT model

The most common way, given the modern world of statistical computing, in which inferences are made under hierarchical Bayesian models, is to obtain samples from the posterior distribution of quantities of interest. The reason why this has become such an attractive way to do inference is multifold; however, before delving into this, we briefly review the most popular and widespread alternatives that are based on maximum marginal likelihood (MML; Bock & Aitken, 1981) and maximium a posteriori (MAP) estimation.

8.3.1. The JML and MML estimation approaches

Binary responses to test items, $y_{ij} = 0$ or 1, lead to an overall likelihood, assuming conditional independence of responses, that is simply the product of a number of Bernoulli trials corresponding to each test item and all respondents. Thus, if we denote the collection of test responses from $i = 1, \ldots,$ I respondents to $j = 1, \ldots, J$ test items as $Y = (y_{ij})$, then the likelihood of observing Y, conditional on the collection $T = (t_{ij})$ – the latent linear predictor of score (given in Section 8.1 above), is

$$L(Y|T) = \prod_{i=1}^{I}\prod_{j=1}^{J} p_{ij}^{y_{ij}}\,(1 - p_{ij})^{y_{ij}} = \prod_{i=1}^{I}\prod_{j=1}^{J}\left(\frac{e^{t_{ij}}}{1+e^{t_{ij}}}\right)^{y_{ij}}\left(\frac{1}{1+e^{t_{ij}}}\right)^{1-y_{ij}}. \tag{8.1}$$

That is, $y_{ij} = 1$ with probability p_{ij} and $y_{ij} = 0$ with probability $1 - p_{ij}$, and the overall probability for all of the data is simply the product of the conditionally independent responses. With $t_{ij} = a_j(\theta_i - b_j - \gamma_{id(j)})$, our 2-PL TRT model, $L(Y|T)$, in (8.1) is a function with unknowns $\theta = (\theta_1, \ldots, \theta_I)$, $a = (a_1, \ldots, a_J)$, $b = (b_1, \ldots, b_J)$, and $\gamma = (\gamma_{1d(1)}, \ldots, \gamma_{Id(J)})$.

As described in Part I, the function in (8.1) could be *jointly* maximized for $\Lambda = (\theta, a, b, \gamma)$ – the so-called joint maximum likelihood (JML) approach; however, this is typically not done for two reasons. First, JML parameter estimates from (8.1) are not consistent (Andersen, 1970) because as $I \to \infty$, the number of parameters, θ_i, increase at the same rate. Second, when one makes inferences, say for the properties of an item (a_j, b_j), one wants to make those statements *unconditionally* on the persons who happened to have taken those items. Therefore, the JML approach is virtually never used in practice, instead people use the MML approach described next. As it is so computationally heavy, it is impractical even for moderately long tests (Wainer, Morgan, & Gustafsson, 1980).

In the MML approach, details of which are presented in Glas, Wainer, and Bradlow (2000), maximization for $\Lambda = (\boldsymbol{\theta}, a, b, \gamma)$ is performed in two stages, which is implemented to rectify the deficiency of the JML approach. In particular, in Stage 1 of the MML approach, one takes (8.1) and "removes" (integrates out) the person-level ability and testlet parameters, (θ, γ), from the model. This is done by integrating (8.1) with respect to the assumed distribution for the $\theta, \theta_i \sim N(0, 1)$, and $\gamma, \gamma_{id(j)} \sim N(0, \sigma_\gamma^2)$, and leads to

$$
\begin{aligned}
L(Y|a, b, \sigma_\gamma^2) &= \iint \prod_{i=1}^{I} \prod_{j=1}^{J} p_{ij}^{y_{ij}} (1 - p_{ij})^{y_{ij}} \, \mathrm{d}F\theta_i \mathrm{d}F\gamma_{id(j)} \\
&= \iint \prod_{i=1}^{I} \prod_{j=1}^{J} \left(\frac{e^{t_{ij}}}{1 + e^{t_{ij}}} \right)^{y_{ij}} \\
&\quad \times \left(\frac{1}{1 + e^{t_{ij}}} \right)^{1 - y_{ij}} N(0, 1) N\left(0, \sigma_\gamma^2\right) \mathrm{d}\theta_i \mathrm{d}\gamma_{id(j)}. \qquad (8.2)
\end{aligned}
$$

Now, with θ_i and $\gamma_{id(j)}$ integrated out, one maximizes (8.2) for the item-level parameters (a, b).

Two things to note about maximizing (8.2) are as follows. First, the integration in (8.2) cannot be done in closed form, as there is no analytic solution to the integrals. This is because the product – Bernoulli likelihood given in (8.1) and the assumed normal distribution priors are not conjugate (i.e., do not belong to the same parametric family). Hence, the integration in (8.2) is commonly performed either through numerical simulation of "plausible values" for the θ_i and $\gamma_{id(j)}$ (Mislevy et al., 1992) or by Gaussian quadrature in which a grid of points (e.g., 5–7) set up on each integration dimension and (8.2), which is a continuous integral, is approximated by a function that is a summation over the finite grid points. Second, even after integrating out θ_i and $\gamma_{id(j)}$, maximization cannot be done directly, as the resulting function's first-order conditions, $\mathrm{d}L(Y|a, b, \sigma_\gamma^2)/\mathrm{d}a = 0$, $\mathrm{d}L(Y|a, b, \sigma_\gamma^2)/\mathrm{d}b = 0$, $\mathrm{d}L(Y|a, b, \sigma_\gamma^2)/\mathrm{d}\sigma_\gamma^2 = 0$, cannot be solved directly. Hence, once the integration is done in (8.2), commonly a numerical optimization technique such as Newton–Raphson is used to obtain the MML estimates, denoted as $\hat{a}, \hat{b}$.

Once $\hat{a}, \hat{b}$ have been obtained, the second stage of MML estimation involves obtaining the maximum likelihood parameter estimates of θ_i and $\gamma_{id(j)}$ but now conditional on the assumed known values of $\hat{a}$, $\hat{b}$ that were obtained in the first step of the MML procedure. That is, the values of $\hat{a}$, $\hat{b}$ are inserted into the function (8.1) and it is then maximized, again commonly using a numerical technique such as Newton–Raphson for the person-level ability parameters. Two things are important. First, from a Bayesian perspective (which is the approach we have taken), the insertion of $\hat{a}$, $\hat{b}$ into (8.1) and treating them as known and fixed is "inappropriate." This is because, as stated above, we have placed hyperpriors on a and b, a log-normal and a normal prior, respectively, and hence "jamming in" point estimates as their values ignore the uncertainty with which they were obtained. If, as in many practical operational settings, the number of respondents I for which $\hat{a}$, $\hat{b}$ are based is very large, then this uncertainty in estimation is likely to be negligible and the uncertainty reduction by using the plug-in method is likely to be small. However, in other operational test settings, the plug-in method has known limitations and should be done with caution. Therefore, this is an issue that should be addressed on a case-by-case basis.

A second comment regarding the plug-in method is that it is a special case of empirical Bayes methods (Efron & Morris, 1973, 1975) or more appropriately Bayes empirical Bayes methods (Morris 1983) in which point estimates of prior parameters are estimated from their marginal distributions and then inserted into the functions. As described later in this chapter, Bayesian methods that allow exact finite sample inferences via posterior sampling do not suffer from these limitations.

In addition to point estimates of the unknown parameters, $\Lambda = (\theta, a, b, \gamma)$, which are generated as a result of the MML method, one could also obtain asymptotic standard errors of parameter estimates that – on the basis of the asymptotic normality of maximum likelihood estimates – are a staple of statistical inference, as significance testing for parameters. For example, $a_j = 0$, is commonly done by computing

$$t(a_j) = \frac{\hat{a}_j}{\mathrm{SE}(\hat{a}_j)} \tag{8.3}$$

and comparing $t(a_j)$ to a set of t-distribution tables. Or, equivalently, one may wish to construct a $100 \times (1 - \alpha)\%$ confidence interval for parameter a_j, given by

$$\hat{a}_j \pm z^*_{\alpha/2}\mathrm{SE}(\hat{a}_j), \tag{8.4}$$

where $z^*_{\alpha/2}$ is the $1 - \alpha/2$ quantile of the standard Gaussian distribution (mean $=$ 0, standard deviation $= 1$).

Finally, we note that the asymptotic standard errors for parameter estimates obtained through an MML procedure can be computed by looking at the diagonal of the negative of the inverse of the Fisher information matrix, evaluated at the MML estimates computed as

$$-\left(\frac{\partial^2}{d\Lambda^2}L(Y|\Lambda = \hat{\Lambda})\right)^{-1}. \tag{8.5}$$

One nice feature of applying an iterative method for computing the MML estimates, such as the Newton–Raphson iterations mentioned previously, is that it requires the computation of (8.5) as part of the algorithm, and hence asymptotic standard errors essentially come for free. One simply evaluates (8.5) at the converged final values (the MML estimates) obtained at the end of the Newton–Raphson procedure.

8.3.2. Maximum a posteriori estimation

The procedures described in Section 8.3.1, which are likelihood-based procedures, can be extended to include the prior distribution information as given in Section 8.2. In particular, (8.2) does not utilize the prior distributions that were posited for the item parameters: $\log(a_j) \sim N(\mu_a, \sigma_a^2)$, $b_j \sim N(\mu_b, \sigma_b^2)$. In maximum a posteriori (MAP) estimation, one simply augments the likelihood given in (8.2) with the prior distributions for the item parameters. Then, since the posterior distribution is proportional to the likelihood multiplied by the prior distribution, we maximize it by using methods identical to MML estimation, and similarly obtain asymptotic standard errors. However, MAP estimation suffers from the same concerns as MML estimation in that plug-in estimates are utilized, and that significance testing as in (8.3) and the construction of confidence intervals as in (8.4) are based on asymptotic normality of the estimators. The fully Bayesian procedures described next do not rely on such an approach.

8.3.3. Bayesian estimation using Markov chain Monte Carlo methods

The model for dichotomous test responses defined by the 2-PL TRT likelihood (given in (8.1)) and its associated priors has a structure that is common to a great deal of contemporary research on hierarchical Bayesian computational methods (Gelman et al., 1995a). In particular, hierarchical Bayesian computational methods focus on obtaining samples from the posterior distribution of model parameters by setting up a Markov chain (Geman & Geman, 1984), which after sufficient sampling, that is, enough iterations of the Markov chain (hereafter referred to as MCMC methods = **M**arkov **c**hain **M**onte **C**arlo methods), converges in distribution to the posterior distribution of interest. There are many reasons why posterior inference using MCMC methods have become popular.

First, one primary advantage of MCMC methods is the ease with which inference can be done once posterior samples have been obtained. Once we have samples from the posterior distribution we have samples! Sample-based inference is then no harder than taking means if we want to estimate the posterior mean of a parameter (which has the nice minimum squared error loss property). Identically, if we want to estimate the p-th quantile of a parameter's distribution, we simply sort the M MCMC draws after convergence from lowest to highest and takes the $p \times M$-th draw (or the nearest integer to $p \times M$). Thus, MCMC methods essentially turn inference into simply adding, counting, and sorting. This is a nice thing and allows different users of the methods to choose whatever inferences they want.

A second reason why MCMC methods have become popular is that they do not rely on asymptotic theory (large-sample theory) to obtain standard errors. Unlike Equation (8.5), which provides standard errors for estimates obtained via maximum likelihood methods (computing the mode), MCMC methods instead permit finite-sample (which is what you have) inferences.

Third, MCMC methods have become increasingly popular due to recent large increases in computing power that have made them feasible. Hence, while early work on MCMC methods (Gelfand & Smith, 1990) provided guidance in cases where the likelihood for the data $L(Y|\Lambda)$ and its associated prior $p(\Lambda|\phi)$ were "compatible" (conjugate – implying easy sampling, to be described shortly), this is no longer a restriction due to the creation of methods such as the Metropolis–Hastings algorithm (Hastings 1970), data augmentation applied to MCMC methods (Tanner & Wong, 1987), Griddy–Gibbs sampling methods (Ritter & Tanner, 1992), adaptive rejection sampling (Gilks, 1992), and slice sampling (Damien, Wakefield, & Walker, 1999) to name a few. Thus, now the researcher can focus on building a model of scientific interest, without the primary concerns resting on the ability to do the computation.

Finally, the popularity of MCMC methods have been greatly increased because of the creation and proliferation of free, user-friendly Bayesian software such as WinBUGS (http://www.mrc-bsu.cam.ac.uk/bugs/). WinBUGS is freeware that enables users to simply specify the hierarchical model that they wish to fit (as in Section 8.1) without having to compute the conditional distributions (see details below) themselves to run the Markov chain. In fact, it is our belief that freeware such as WinBUGS has had more impact on the popularity and increased attention to Bayesian methods than all of the journal articles combined. We as researchers have impact on practice when we provide software that people can use to implement these methods. We should also note, in the spirit of WinBUGS, that we (the authors) have created a freeware program, SCORIGHT 3.0, that enables inferences for IRT models using MCMC methods. The details of the current implementation of the program are saved for Chapter 12, until the complete Bayesian testlet model has been laid out in full generality. Now, onto the details of how MCMC sampling works in general, and in particular, for the 2-PL Bayesian IRT model laid out in Section 8.1.

Next, let us examine the algorithmic way in which MCMC methods work. Let Λ denote the parameters of the model. This includes both the parameters of the likelihood, priors, and hyperpriors (if they exist). For example, in the Bayesian 2-PL testlet model specified in Section 8.1, Λ includes the 2-PL item parameters $(a_1, \ldots, a_J)$, $(b_1, \ldots, b_J)$, the ability and testlet parameters $(\theta_1, \ldots, \theta_I)$, $(\gamma_{1d(1)}, \ldots, \gamma_{Id(J)})$, and the parameters that govern their distributions (μ_a, σ_a^2), (μ_b, σ_b^2), and σ_γ^2. An MCMC sampler then proceeds through the following steps. We note that there is not just one MCMC sampler that is possible; in particular, how one cycles through the parameters (STEP 2 below) does not only have a single solution, and, in fact, any setup that is space filling (Liu, 1993) is allowable.

[STEP 1] Select an initial starting vector, $\Lambda = \Lambda^{(t=0)}$, where t denotes the iteration number. Set $t = 0$. Our experience indicates that the results from programs such as MULTILOG and BILOG provide excellent starting values for MCMC samplers. In fact, using their final results as starting values can speed up convergence to the ultimate stationary distribution significantly.

[STEP 2] Select some subset of the parameters, λ_1, and draw updated values $\lambda_1^{(t+1)}$ from its full conditional distribution $p(\lambda_1 | Y, \Lambda_{-\lambda_1}^{(t)})$, where $\Lambda_{-\lambda_1}^{(t)}$ denotes the entire parameter vector Λ excluding parameters λ_1, evaluated at its t-th value, and Y denotes the observed test data as before.

[STEP 3] Select some subset of the parameters λ_2, and draw updated values $\lambda_2^{(t+1)}$ from its full conditional distribution $p(\lambda_2 | Y, \Lambda_{-\lambda_1,-\lambda_2}^{(t)}, \lambda_1^{(t+1)})$, where $\Lambda_{-\lambda_1,-\lambda_2}^{(t)}$ denotes the set of all parameters excluding λ_1 and λ_2, where λ_2 is evaluated at its t-th value, and $\lambda_1^{(t+1)}$ is the updated value of λ_1 obtained in STEP 2.

[STEP 4] Sample $\Lambda_{-\lambda_1,-\lambda_2}^{(t+1)}$ from its full conditional distribution $p(\Lambda_{-\lambda_1,-\lambda_2} | Y, \lambda_1^{(t+1)}, \lambda_2^{(t+1)})$. Let $t = t + 1$.

[STEP 5] If $t \leq M$ (a prespecified value), go to STEP 2, if not, then stop.

A number of important implementation "choices" are not mentioned in STEPS 1–5 above, yet may significantly impact the results of the Bayesian 2-PL testlet model MCMC sampler. We provide a discussion of many of these issues next. For a more complete discussion of implementation issues see Gelman et al. (1995). These same implementation issues are common and are relevant to the discussion in Chapters 9 to 11 as well but are only addressed here.

As starting values are the point at which all MCMC samplers originate, we shall begin there too. Obtaining good starting values for some MCMC samplers can be crucial for models that may be multimodal or are high-dimensional. The reason for this is that convergence of Markov chains to their stationary distribution is anything but an exact science, and not to be taken lightly; rather, there are a number of significance tests that are used to indicate convergence (Geyer, 1994; Gelman & Rubin, 1992; Sinharay 2004). Fortunately, the class of IRT models that we are discussing in this chapter, the 2-PL testlet model, is well behaved, and hence convergence of the MCMC sampler can be achieved fairly quickly (for typical data sets within 10,000 iterations). In addition, as mentioned in STEP 1, other existing software (albeit not testlet based) can provide excellent starting values.

Another important implementation issue, as exemplified by $\lambda_1, \lambda_2, \Lambda_{-\lambda_1,-\lambda_2}$, is how to break up the parameter vector, Λ, in order to implement the MCMC sampler. In the case of IRT models, this is important to state explicitly, for even though STEPS 2 to 4 above can be simply stated as "just sample from their full conditional posterior distributions," such advice is not trivial to carry out for IRT models because the form of the likelihood given in (8.1), is nonconjugate with the Gaussian priors that are commonly assumed for its parameters. Hence, the reason that Equation (8.2) cannot be computed in closed form is for the same reason that sampling from the full conditional distributions cannot be done directly. However, even though much research has been done over the past fifteen years that allows for sampling from these distributions, care must be taken. In particular, while there is a natural grouping of parameters in the Bayesian 2-PL testlet model (a, b, θ, γ), the methods needed to sample these entire vectors, say the Ja's all at once, would likely have poor MCMC mixing properties; the Markov chain might get stuck at specific values, or might just move very slowly through the parameter space. Metaphorically, one can imagine a topologist trying to map out the posterior surface, but getting stuck in a valley, or somewhere else that has low likelihood in comparison with the area near the mode of the posterior surface.

The most common method that would be used to sample from the full conditional distribution is the Metropolis–Hastings algorithm. In this algorithm, one selects a potential value λ^* from a proposal distribution, $g(\lambda)$, and then either "stays" by setting $\lambda^{(t+1)}$ equal to the previous value $\lambda^{(t)}$ or sets $\lambda^{(t+1)} = \lambda^*$, with probability given by the ratio

$$\min\left(\frac{p(\lambda^*|Y, \Lambda_{-\lambda}^{(t)})g(\lambda^*)}{p(\lambda^{(t)}|Y, \Lambda_{-\lambda}^{(t)})g(\lambda^{(t)})}, 1\right). \tag{8.6}$$

For high-dimensional vectors, the probability of selecting a "good" candidate vector, λ^*, is low and hence the MCMC sampler gets stuck at values near or exactly equal to $\lambda^{(t)}$. To help ameliorate this problem, (not the only way) our experience has shown that sampling from the full conditional distribution of Λ

one-dimension at a time (i.e., a_3 or θ_{16}, etc.) makes the acceptance rate of the MH algorithm reasonable and allows one to control well the mixing properties of the MCMC sampler (Gelman, Roberts, & Gilks, 1995). Hence, in SCORIGHT 3.0, our implementation of the Bayesian 2-PL testlet model, each of the parameters from Λ is sampled one at a time. Further empirical study and/or theoretical research is needed to assess whether this is the most efficient implementation; nevertheless, in our experience, this one has always worked so far and requires the least amount of fine tuning.

One final aspect of the implementation of an MCMC sampler that we discuss is the choice of the number of iterations, that is, M. In essence, M is composed of two parts, M' and $M - M'$. The first M' iterations of the MCMC sampler are called a "burn-in" period in which, using the same exploring topologist analogy who may have started off far away from the significant parts of the posterior surface (due to unfortunate starting values), is "walking toward" the area of the posterior that has significant probability. The remaining $M - M'$ iterations are utilized to estimate the posterior quantities of interest when the MCMC sampler has reached stationarity and hence is now sampling from the stationary distribution of the parameters. In choosing $M - M'$, one balances off computation time with Monte Carlo simulation error.

The most widely used convergence diagnostics, that is in selecting M', essentially do an F-test looking at the ratio of within-to between-chain variation, and then assess the Markov chain that has converged when the between-chain variation is small in comparison with the within-chain variation. This requires the analyst to (a) run multiple (say Q) MCMC samplers, as in STEPS 1–5 above, so that an F-test of between-to within-chain variation can be assessed and (b) after convergence, take the $Q \times (M - M')$ draws and use those for posterior inference. Commonly chosen values of Q are 3–5, which are large enough so that convergence can be assessed yet small enough to not make computational time infeasible (computation time goes up at least linearly in the number of chains implemented).

With these details laid out, the next section provides a simulation of the efficacy of MCMC samplers to give valid inferences. These results are taken from Bradlow, Wainer, and Wang (1999).

8.4. A demonstration of the efficacy of the Bayesian 2-PL testlet model

When the Bayesian 2-PL testlet model (given above) was introduced in Bradlow, Wainer, and Wang (1999), it had not been verified that an MCMC approach by obtaining posterior samples would provide valid inferences under this model. To assess this, an extensive set of simulations was run. A summary of their results are given below. Each of the simulations was run assuming $I = 1{,}000$ examinees taking a test composed of $J = 60$ items. Each MCMC simulation condition was run for $Q = 3$ independent chains started from overdispersed starting values, Each simulation was diagnosed to have converged after approximately $M' = 1{,}000$ iterations: M was set to 2,000. Hence, there were $Q \times (M - M') = 3{,}000$ draws that were used to provide inferences for each simulation. Inferences and the subsequent results were obtained by computing posterior means for the parameters of interest and comparing them, in various ways, to the true values that were used to simulate the data.

In each simulation, except Simulation 1, which acted as a control condition, 30 of the 60 items were independent items; the remaining 30 were nested within testlets. The two important variables to control (across simulations) were the number of items per testlet (varied between 5 and 10, that is, six testlets of Size 5 making up the 30 testlet items, or three testlets of Size 10), and the amount of testlet variance. The reason the number of items per testlet was varied was that it was of interest to understand how the MCMC algorithm would perform where the testlet size was more moderate (say five items per testlet) than those of large size (10 items per testlet), and hence the amount of information was sparse. We note that further testing suggests that even smaller testlet sizes than 5 yield adequate results. Second, the amount of testlet variance, σ_γ^2, was also considered critical, as it was important both to understand how strong the testlet "signal" had to be before the model would pick it up and to see the effects of testlet variance when it is ignored (i.e., fitting a model in which σ_γ^2 was erroneously assumed to be 0). To norm the size of the testlet variance, we look at the ratio of testlet variance σ_γ^2 to the variance of θ_i, which is normalized to 1. In particular, we chose values where the testlet variance would be half the size of the variation associated with the examinee abilities, equal to that variation, and twice that size. Empirical work has demonstrated that these are plausible values.

The following is a synopsis of the simulation results for seven simulation conditions created from a 2×3 factorial design plus the control condition:

(S1) Control condition: no testlet effect, all independent items;
(S2) Testlet size $= 5$, $\sigma_\gamma^2 = 0.5$;
(S3) Testlet size $= 5$, $\sigma_\gamma^2 = 1$;
(S4) Testlet size $= 5$, $\sigma_\gamma^2 = 2$;
(S5) Testlet size $= 10$, $\sigma_\gamma^2 = 0.5$;
(S6) Testlet size $= 10$, $\sigma_\gamma^2 = 1$; and
(S7) Testlet size $= 10$, $\sigma_\gamma^2 = 2$.

Table 8.1. *Simulation results of the mean absolute error of prediction for the Bayesian 2-PL testlet model*

Testlet size	Var(γ)	Condition	BILOG θ	BILOG a	BILOG b	MCMC θ	MCMC a	MCMC b	MCMC γ θ	MCMC γ a	MCMC γ b
1	0.00	S1	0.20	0.06	0.07	0.20	0.06	0.06	0.20	0.06	0.06
5	0.50	S2	0.22	0.07	0.09	0.22	0.07	0.08	0.22	0.05	0.07
	1.00	S3	0.25	0.12	0.11	0.26	0.13	0.13	0.24	0.09	0.13
	2.00	S4	0.26	0.13	0.12	0.26	0.13	0.12	0.24	0.06	0.12
10	0.50	S5	0.23	0.06	0.09	0.23	0.05	0.08	0.22	0.06	0.07
	1.00	S6	0.28	0.09	0.11	0.28	0.08	0.11	0.23	0.06	0.07
	2.00	S7	0.25	0.11	0.11	0.35	0.12	0.12	0.25	0.07	0.11
		Mean	**0.26**	**0.09**	**0.10**	**0.26**	**0.09**	**0.10**	**0.23**	**0.06**	**0.09**

Our findings can be summarized in two tables. The first, Table 8.1, contains the mean absolute error of prediction for (θ, a, b). For all simulation conditions, we compare the results of the MCMC method with the testlet effect included, denoted MCMCγ; the MCMC method with the testlet effect erroneously (except for Simulation condition 1) ignored, denoted MCMC; and BILOG. The second table, Table 8.2, contains a comparison of 95% coverage probabilities for θ for both MCMCγ and MCMC.

On the basis of the results in Table 8.1, we note first that the mean absolute errors for both the MCMC and MCMCγ approaches demonstrate the efficacy of the model and computational approach. The posterior means for the MCMC approaches are consistent with extant established software BILOG. And second, for S1, all three approaches yield the same results for the mean absolute error. This is not surprising given the fact that there is no testlet effect in these data and all three models should be "shooting at the same target." However, as expected, it is still crucial that the MCMCγ approach does not find a testlet effect when in fact there are none. Third, as the testlet variance increases, the mean absolute error increases for all the three methods; albeit more dramatically for the two methods (BILOG and MCMC) that do not model the testlet variance. Fourth, as testlet size increases, the mean absolute error for θ also increases for the nontestlet methods. This makes sense, as there is less information available for estimating θ in testlets of larger size, for the larger number of items provides the opportunity for greater redundancy among the items. Finally, the MCMCγ method outperforms the other two methods and, as expected, the results for MCMC and BILOG are the same.

Table 8.2. *Ninety-five percent coverage probabilities for θ*

Testlet size	Var(γ)	Condition	MCMC	MCMCγ
1	0.00	S1	0.94	0.94
	0.50	S2	0.90	0.94
5	1.00	S3	0.92	0.96
	2.00	S4	0.87	0.93
	0.50	S5	0.88	0.95
10	1.00	S6	0.84	0.96
	2.00	S7	0.75	0.95
		Mean	**0.87**	**0.95**

Table 8.2 shows the coverage probabilities (i.e., fraction of posterior intervals that cover the true values of θ). Since, these intervals are constructed to be 95% posterior intervals, they should have at least 95% coverage. As the results indicate, the MCMCγ method has the correct nominal coverage, whereas the MCMC method has severe undercoverage. The degree of undercoverage increases as the testlet variance increases, as well as when the number of items per testlet increases. These results are consistent with the idea that when the amount of information is falsely overestimated, as would be done by ignoring the excess within-testlet covariance, then posterior intervals are too narrow, and hence do not have the appropriate coverage. In fact, it is even "worse than that" in that you now have overconfidence in your belief about coverage.

In summary, the Bayesian 2-PL testlet model, implemented via MCMC methods, is able to account for the testlet dependence in an IRT model for 2-PL binary items. We describe next how this model and subsequent methods have been generalized.

8.5. A roadmap for the remaining computational chapters

The Bayesian 2-PL model described in this chapter, while important in linking traditional IRT models to other works that have identified deficiencies in the assumption of unidimensional conditional independence (Hedeker & Gibbons, 1994; Zhang & Stout, 1999), also has a number of limitations that require generalization. In particular, the model given in Section 8.1 is limited by:

1. Constant testlet variance. As implemented in Section 8.2, the assumed model for testlet effects, $\gamma_{id(j)} \sim N(0, \sigma_\gamma^2)$, assumes a constant variance, σ_γ^2, across testlets. This is unlikely to hold in practice, as some testlets

exhibit greater dependence than do others. Our extended model, to be discussed in Chapter 9, allows for this.

2. It only utilizes the 2-PL model. While the 2-PL model has significant operational use, it is limited in its application to multiple-choice items because it ignores the often likely possibility of guessing. That is, examinees who are "infinitely unable" ($\theta = -\infty$) have, under the 2-PL model, no chance of getting the item correct. But with a K-choice item, random guessing would yield a probability of $1/K$ correct in the long run. Hence, in Chapter 9, we also discuss the extension of the Bayesian 2-PL testlet model to the Bayesian 3-PL testlet model.
3. The model considers only binary data. As richer forms of assessment (essays, portfolios, etc.) exist and become more commonplace, data in the form of ordinal scores are much more likely to exist. In fact, many educational tests are a blend of multiple-choice sections and ordinally scored parts. In Chapter 10, we extend the Bayesian testlet model to include data sets that are a mixture of dichotomous and ordinal data.
4. The model ignores potentially available covariate information. In many real testing situations, one may wish not only to account for testlet dependence, as $\gamma_{id(j)}$ allows one to do, but also to explain the sources of it. For example, do passages about certain topics have more testlet dependence than others? Do passages with more words have more testlet dependence than others? In fact, covariate explanations are not limited to testlet effects, rather we might be interested in understanding how item difficulties or discriminations vary by item type. Or, we may be interested in understanding how person-level abilities vary by characteristics of the individual. In Chapter 11, we build a unified Bayesian model that allows for covariates to be brought in to explain Λ.

Each of these additions has been implemented and are all part of SCORIGHT 3.0 (Wang, Bradlow, & Wainer, 2004).

Questions

1. Explain how the 2-PL TRT model is related to a factor analysis model.
2. Explain how the 2-PL TRT model leads to a block-diagonal covariance structure for the outcomes Y.
3. Describe the principles by which one would select a sampling distribution g when implementing a Metropolis–Hastings algorithm.

4. What is the rationale for running multiple MCMC chains and diagnosing convergence using an F-test versus running one long chain? What is a rationale for running one long chain versus running multiple chains of shorter length?
5. Describe three key limitations of the 2-PL TRT model described in this chapter and how you might address them.
6. Provide an example of two distributions that are conjugate. Furthermore, derive the posterior distribution, demonstrating that it is of the same family as the prior. Finally, describe why the Bayesian 2-PL TRT model is not a conjugate model.
7. Writing code in WinBUGS, or some related package, is important in being able to implement MCMC methods. Write code in WinBUGS that would allow you to fit the Bayesian 2-PL TRT model.
8. Explain the similarities and differences between MML, JML, maximum a posteriori, and Bayesian inference using posterior samples in terms of the computational approach, parameter estimates obtained, and the way in which standard errors are computed.

References

Andersen, E. B. (1970). Asymptotic properties of conditional maximum likelihood CMLE estimators. *Journal of the Royal Statistical Society, Series B, 32*, 283–301.

Bock, R. D., & Aitken, M. (1981). Marginal maximum likelihood estimation of item parameters. An application of an EM algorithm. *Psychometrika, 46*, 443–459.

Bradlow, E. T., Wainer, H., & Wang, X. (1999). A Bayesian random effects model for testlets. *Psychometrika, 64*, 153–168.

Damien, P., Wakefield, J. C., & Walker, S. G. (1999). Gibbs sampling for Bayesian nonconjugate and hierarchical models by using auxiliary variables. *Journal of the Royal Statistical Society, B Statistical Methodology, 61*, 331–344.

Efron, B., & Morris, C. (1973). Stein's estimation rule and its competitors – An empirical Bayes approach. *Journal of the American Statistical Association, 68*, 117–130.

Efron, B., & Morris, C. (1975). Data analysis using Stein's estimator and its generalizations. *Journal of the American Statistical Association, 70*, 311–319.

Gelfand, A. E., & Smith, A. F. M. (1990). Sampling-based approaches to calculating marginal densities. *Journal of the American Statistical Association, 85*, 398–409.

Gelman, A., Carlin, J. B., Stern, H. S., & Rubin, D.B. (1995). *Bayesian data analysis*. London: Chapman & Hall.

Gelman, A., Roberts, G., & Gilks, W. (1995). Efficient Metropolis jumping rules. In *Bayesian statistics 5*. New York: Oxford University Press.

Gelman, A., & Rubin, D. B. (1992). Inference from iterative simulation using multiple sequences. *Statistical Science, 7*, 457–472.

Geman S., & Geman, D. (1984). Stochastic relaxation, Gibbs distributions and the Bayesian restoration of images. *IEEE Transactions on Pattern Analysis and Machine Intelligence, 6*, 721–741.

Geyer, C. J. (1994). On the convergence of Monte Carlo maximum likelihood calculations. *Journal of the Royal Statistical Society, Series B, 56*, 261–274.

Gilks, W. R. (1992). Derivative-free adaptive rejection sampling for Gibbs sampling. In J. M. Bernardo, J. O. Berger, A. P. Dawid, & A. F. M. Smith (Eds.) *Bayesian Statistics, 4* (pp. 169–194). Oxford, UK: Clarendon Press.

Glas, C. A. W., Wainer, H., & Bradlow, E. T. (2000). Maximum marginal likelihood and expected a posteriori estimation in testlet-based adaptive testing. In W. J. van der Linden & C. A. W. Glas (Eds.), *Computerized adaptive testing, theory and practice* (pp. 271–288). Boston, MA: Kluwer-Nijhoff.

Hastings, R. (1970). Monte Carlo sampling methods using Markov chains and their applications. *Biometrika, 54*, 93–108.

Hedeker, D., & Gibbons R. D. (1994). A random-effects ordinal regression model for multilevel analysis. *Biometrics, 50*, 933–944.

Liu, J. (1993). The collapsed Gibbs sampler with applications to a gene regulation problem. *Journal of the American Statistical Association, 89*, 958–966.

Mislevy, R. J., Beaton, A., Kaplan, B. A., & Sheehan, K. (1992). Estimating population characteristics from sparse matrix samples of item responses. *Journal of Educational Measurement, 29*(2), 133–161.

Morris, C. N. (1983). Parametric empirical Bayes inference: Theory and applications with discussion. *Journal of the American Statistical Association, 78*, 47–65.

Ritter, C., & Tanner, M. A. (1992). Facilitating the Gibbs sampler: The Gibbs stopper and the Griddy-Gibbs sampler. *Journal of the American Statistical Association, 87*, 861–868.

Sinharay, S. (2004). Experiences with Markov chain Monte Carlo convergence assessment in two psychometric examples. *Journal of Educational and Behavioral Statistics, 29*(4), 461–488.

Tanner, M. A., & Wong, W. H. (1987). The calculation of posterior distributions by data augmentation (with discussion). *Journal of the American Statistical Association, 82*, 528–550.

Thissen, D., Chen, W.-H., & Bock, R. D. (n.d.). MULTILOG 7 – Analysis of multiple-category response data. Retrieved from http://www.assess.com/Software/MULTILOG.htm

Wainer, H., Bradlow, E. T., & Du, Z. (2000). Testlet response theory: An analog for the 3-PL useful in testlet-based adaptive testing. In W. J. van der Linden, C. A. W. Glas (Eds.), *Computerized adaptive testing, theory and practice* (pp. 245–270) Boston, MA: Kluwer-Nijhoff.

Wainer, H., Morgan, A., & Gustafsson, J.-E. (1980). A review of estimation procedures for the Rasch model with an eye toward longish tests. *Journal of Educational Statistics, 5*, 35–64.

Wang, X. A., Bradlow, E. T., & Wainer, H. (2004). A user's guide for SCORIGHT (version 3.0): A computer program built for scoring test built of testlets including a module for covariate analysis (ETS Research Report RR 04-49). Princeton, NJ: Educational Testing Service.

WinBUGS, Retrieved from http://www.mrc-bsu.cam.ac.uk/bugs/welcome.shtml

Zhang, J., & Stout, W. F. (1999). The theoretical DETECT index of dimensionality and its application to approximate simple structure. *Psychometrika, 64*, 213–249.

9

The 3-PL Bayesian testlet model

Two of the limitations of the model described fully in Chapter 8 that are addressed here are (1) the ability of the model to handle binary data in which guessing may occur (or at least may significantly affect the probability of someone getting an item correct) and (2) the idea that testlets may exhibit markedly differing amounts of testlet dependence, and hence a single testlet variance parameter, σ_γ^2, is insufficient. As we demonstrate, both of these additions to the model extend it to be more realistic, have computational implications for it, and have a large impact on practice, as both effects exist empirically to a large and meaningful extent.

9.1. The 3-PL Bayesian testlet model

To describe issue (1) in more detail, consider the solid line 2-PL item characteristic curve (ICC), shown in Figure 9.1, where proficiency $= \theta$ is on the x-axis and the probability correct, $P(y_{ij} = 1)$, is on the y-axis, where item discrimination, a, is set to 1, and item difficulty, b, is set to 0.

Imagine an individual whose proficiency is $\theta = -4$, that is, 4 standard deviations below the mean ability (remember $\theta_i \sim N(0, 1)$). Under the 2-PL model, she is modeled as having essentially 0 probability of getting a given item correct. However, imagine that the item under consideration is a multiple-choice item and has R (say five) possible options. Then, by random guessing, this individual would have $1/R$ probability of getting the item correct, and hence the prediction of the 2-PL model is bound to be inaccurate.

To ameliorate this problem, consider instead the dotted line ICC in Figure 9.1, which is the ICC from the standard 3-PL model (Birnbaum, 1968) – that is, no testlet effect – which adds an additional parameter, commonly denoted as c_j and

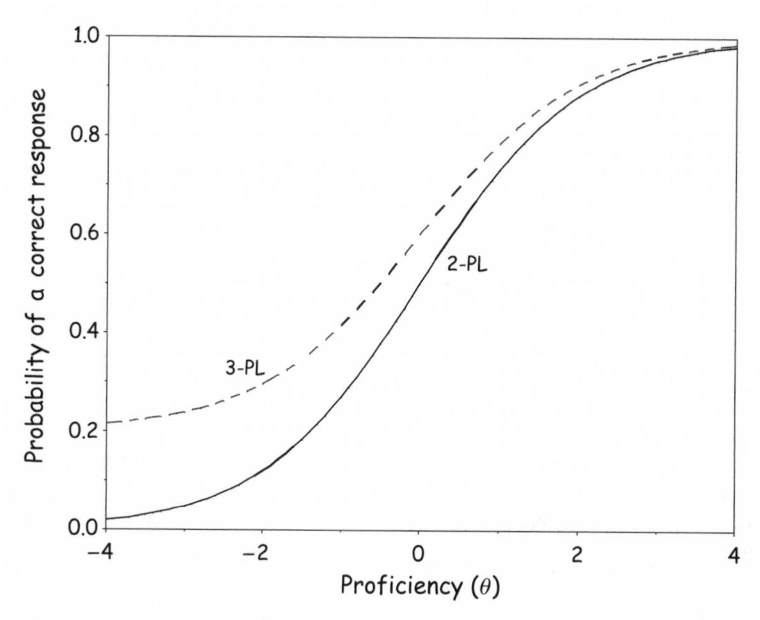

Figure 9.1. ICCs of the 2-PL model (solid line) and the 3-PL model (dashed line) for an item with a = 1 and b = 0.

called the "guessing parameter" of item j, where $P(y_{ij} = 1)$ is given by

$$P(y_{ij} = 1) = c_j + (1 - c_j)\text{logit}^{-1}(a_j(\theta_i - b_j)). \tag{9.1}$$

As before with the 2-PL model, a_j and b_j are the item discrimination and difficulty parameter, θ_i is the person-level proficiency parameter, and $\text{logit}^{-1}(x) = \exp(x)/(1 + \exp(x))$. Now note that when $\theta_i = -4$, or more generally goes to $-\infty$, instead of $P(y_{ij} = 1) = 0$, we have $P(y_{ij} = 1) = c_j$, where the value of c_j in Figure 9.1 was chosen as 0.2 (i.e., $1/R = 1/5$), although it is typically estimated.

With this 3-PL model in hand, one might ask the question "If there is this more general model, why would one ever fit the 2-PL model?" This is a legitimate question to ask as the 2-PL model is a special case of the 3-PL model when $c_j = 0$. Hence, one argument would be to fit the more general 3-PL model, and if c_j appears to be equal to 0, the data would suggest $c_j = 0$ as a plausible value.

Yet, despite this increase in generality, the 3-PL model is not fit in all operational test settings because when the binary data IRT model is augmented with the guessing parameter, its estimation becomes more difficult in the range of the data due to weak identifiability, that is, presence of multiple different triplets of (a_j, b_j, c_j) for which the ICCs are close. Empirical studies have demonstrated this difficulty when implementing the model, and hence when the 3-PL model is used, it should be used with caution. We agree with this tenet, and extensive simulations support the wisdom of caution in drawing inferences from the values of these parameters when sample sizes are modest.

The model in (9.1), while extending the 2-PL model of Chapter 8 to allow for guessing, does not incorporate the extra dependence for items nested within testlets that is likely to be seen in practice. Thus, instead, the 3-PL testlet model is given by its analogous 3-PL TRT version:

$$P(y_{ij} = 1) = c_j + (1 - c_j)\text{logit}^{-1}(a_j(\theta_i - b_j - \gamma_{id(j)})), \qquad (9.2)$$

where $\gamma_{id(j)}$ again is the testlet effect when respondent i is confronted with item j nested within testlet $d(j)$. Although this aspect of the 3-PL model is directly extended from Chapter 8 and the 2-PL model, what we describe in Section 9.2 is an important extension in which the model reflects the possibility that differing testlets may have different amounts of local dependence.

The prior distributions, $\theta_i \sim N(0, 1)$, $\log(a_j) \sim N(\mu_a, \sigma_a^2)$, and $b_j \sim N(\mu_b, \sigma_b^2)$, are kept the same as previously specified. Since an additional parameter, c_j, has been added to the model, one additional prior specification is needed, that is, for c_j. Since c_j is bounded between 0 and 1, we model instead $\text{logit}(c_j) = \log(c_j/(1 - c_j))$, which is defined on the real number line as coming from $N(\mu_c, \sigma_c^2)$, consistent with the other item parameters. In addition, weak but slightly informative conjugate hyperpriors are put on μ_a, σ_a^2, μ_b, σ_b^2, μ_c, and σ_c^2 as in Chapter 8 to complete the Bayesian model specification.

9.2. Varying local dependence within testlets

While the model laid out in Chapter 8 does *not* restrict the testlet effect to be the same for every testlet, that is, it has $\gamma_{id(j)}$ and not γ_i, the prior distribution that was placed on the testlet effect, $\gamma_{id(j)} \sim N(0, \sigma_\gamma^2)$, is overly restrictive. In essence, σ_γ^2, and in particular the ratio of σ_γ^2 to $\sigma_\theta^2 = 1$, $\sigma_{a,}^2$ and σ_b^2 determine the degree of extra dependence due to the testlet effect, but this ratio under this restrictive model is the *same* for every testlet. Thus, to generalize this aspect of the model, as well as generalizing the model from 2-PL to 3-PL, the prior distribution for the testlet effects is changed to $\gamma_{id(j)} \sim N(0, \sigma_{d(j)}^2)$, where now there is a testlet-specific variance component.

As we demonstrate in Section 9.3 through simulation, and in Section 9.4 in which we present some summary results from an analysis of data from an administration of the Scholastic Assessment Test (SAT) and Graduate Record Examination (GRE) data, this is an important extension of the model.

Before going into the results, another aspect of the Bayesian 3-PL model that is worth discussing briefly are its computation features. While most aspects of the computation are common to that of the 2-PL model (it can be done by utilizing an MCMC sampler with a Metropolis–Hastings step for those parameters that are part of the IRT model), this model is more weakly identified; the data have less to say about the model parameters' exact values. This will manifest itself in two major ways. First, one should expect to have to run this model for a greater number of iterations, M', for it to achieve convergence. A more weakly identified model typically has a flatter likelihood surface, and hence moving around the surface (i.e., having the Markov chain mix properly) is a more challenging task. However, as long as one obtains convergence diagnostics using multiple chains, obtained from differing starting values (Gelman & Rubin, 1992), this is not a difficult issue to overcome. Second, we expect that parameter recovery is likely to be degraded for the 3-PL model in comparison with the 2-PL or 1-PL model.

Another computational change between the 2-PL and 3-PL models is that the former can be implemented using methods in which the latent linear predictor of score, $t_{ij} = a_j(\theta_i - b_j - \gamma_{id(j)})$, is drawn from its full conditional distribution. All other parameters are drawn from $p(\lambda|\Lambda_{-\lambda}, Y, T)$, now conditional on $T =$ the matrix of t_{ij}, which can then be done in closed form. This nice computational feature of the 2-PL model, is not as easily done for the 3-PL model, and hence a Metropolis–Hastings procedure for the 3-PL model is more common. A computational comparison of the Metropolis–Hastings and data augmentation approaches was done in Bradlow (2000) for the Rasch model, and could easily be extended to more complex IRT models. It is an area for future empirical study to determine which computational approach is more efficient – a topic of particular interest as data sets grow in size.

We next turn to a summary of some simulation results.

9.3. Simulation results for the Bayesian 3-PL testlet model

This simulation was implemented for three major purposes.

1. To test whether MCMC methods would be able to recover parameters used for simulating data for the Bayesian 3-PL testlet model.
2. To compare the results of four different methods:
   i. MML implemented using BILOG;

ii. MCMC = a Markov chain Monte Carlo 3-PL model without testlet effects;
iii. MCMCγ = a Markov chain Monte Carlo 3-PL model with a common testlet effect variance, σ_{γ}^2, and
iv. MCMC$_\text{d}$ = a Markov chain Monte Carlo 3-PL model with a different testlet variance by testlet, $\sigma_{d(j)}^2$.

3. To simulate data with and without testlet effects, both nonvarying and varying by testlet, and see what impact that would have on each of the four models', (i–iv), ability to recover the true parameter values.

In some sense, even though we had some answers to the question posed in Purpose 1 for the Bayesian 2-PL testlet model (Chapter 8), the "weak identifiability" of the Bayesian 3-PL testlet model made us think that parameter recovery for this model was not to be taken as given. In addition, as MCMC$_d$ compared to MCMCγ was the main additional contribution of this work, we were particularly interested in seeing what would happen if differential testlet effects existed, but only one fits MCMCγ. Also, we were interested in whether when we ran MCMC$_d$ on different real data sets would there be significant differences in a Markov chain Monte Carlo 3-PL model with a common testlet effect variance, $\sigma_{d(j)}^2$, across testlets, and if so, to what magnitude. Hence, our interest here were both theoretical and empirical.

The simulation design utilized was as follows. Each data set consisted of the binary scores of 1,000 simulees responding to a test of 70 multiple-choice items. Each of the 70-item tests consisted of 30 independent items, and four testlets containing 10 items each (a test of realistic and practical length). In fact, much of the simulation design described here was done to mimic the real SAT and GRE data to be described in the next section. Each of the simulated data sets varied according to the amount and consistency of the values of $\sigma_{d(j)}^2$, across the four 10-item testlets, as follows:

Table 9.1. *Simulation design for the Bayesian 3-PL testlet model, where j-th entries in the table are the values of* $\sigma_{d(j)}^2$

Data set	Testlet 1	Testlet 2	Testlet 3	Testlet 4
1 (No effect)	0.00	0.00	0.00	0.00
2 (Equal effects)	0.80	0.80	0.80	0.80
3 (Unequal effects)	0.25	0.50	1.00	2.00

Hence, data set 1 contained no testlet effects, and allows us to test whether each of the models can appropriately collapse down to a simpler model when

Table 9.2. *Correlations of estimated parameters with true values for the Bayesian 3-PL testlet model*

	θ	a	b	c	Mean
No effect					
MML	0.96	0.84	0.98	0.56	0.84
MCMC	0.96	0.93	0.99	0.64	0.88
MCMCγ	0.96	0.93	0.99	0.63	0.88
MCMC$_d$	0.96	0.93	0.99	0.66	0.89
Equal effects					
MML	0.92	0.77	0.98	0.64	0.83
MCMC	0.92	0.80	0.98	0.65	0.84
MCMCγ	0.93	0.87	0.99	0.69	0.87
MCMC$_d$	0.93	0.86	0.99	0.70	0.87
Unequal effects					
MML	0.91	0.76	0.97	0.57	0.80
MCMC	0.91	0.80	0.97	0.60	0.82
MCMCγ	0.92	0.75	0.98	0.68	0.83
MCMC$_d$	0.93	0.83	0.99	0.68	0.86
Mean	**0.93**	**0.84**	**0.98**	**0.64**	**0.85**

increased local dependence is not an issue. Data set 2 allows us to look at both the impact of testlet variance when it is ignored for the 3-PL model (using both the MML and MCMC methods), and also whether the MCMC$_d$ method would simplify down to the MCMCγ method when $\sigma^2_{d(j)} = \sigma^2_\gamma$, the same for all testlets $d(j) = 1, 2, 3, 4$. Finally, data set 3 is the richest structure in which the testlet effect variance varies over testlets, and we would expect that all of the estimation methods except for MCMC$_d$ would suffer in terms of parameter recovery. Thus, our overall design was a 3 (data set) $\times$ 4 (estimation method = MML, MCMC, MCMCγ, and MCMC$_d$) design.

To complete a description of the model specification, the remaining parameters were drawn from Gaussian distributions such that $\theta_i \sim N(0, 1)$, $a_j \sim N(0.8, 0.2^2)$, $b_j \sim N(0, 1)$, and $c_j \sim N(0.2, 0.03^2)$.

The main set of summary simulation results is presented in Table 9.2 in which we show the correlation between each of the parameter sets estimated values (modal values in the case of MML and posterior means in the case of the MCMC, MCMCγ, and MCMC$_d$ methods) and the true simulated values. We do this for each of the 3×4 conditions and for each of (θ, a, b, c) parameters.

These simulation results provide a number of confirmatory and interesting findings. First, overall, the parameter recovery is quite good. Except for the c parameter, all correlations are high (above 0.9 for θ and b as is normally observed), fairly high for a (mostly between 0.8 and 0.9), and moderate (and at expected levels around 0.6–0.7) for c. Of course, the correlation may not tell the entire story, as there could be systematic biases that still make the actual prediction errors poor. However, we note, in tables not shown here but discussed in Wainer, Bradlow, and Du (2000), that this is not the case and the estimated parameters are adequately recovering the simulated true values. Second, we note that the c parameters are predicted more poorly by all models, which is as expected and has been noted in the literature. Third, when testlet effects do not exist, all of the models (except for MML) perform approximately the same: however, when testlet variance begins to play a role, the fit of models, which ignore that local dependence (MML and MCMC), degrade. Finally, when the testlet variance has unequal effects, the best fitting model is MCMC_d. One might suggest that the results are very close between MCMC_d and MCMCγ, which is true in an absolute sense: however, when we note that the results for MCMCγ are already close to 1, then a better measure of the improved efficacy of MCMC_d than for MCMCγ is the percentage of remaining correlation left that is explained. When using this metric, we see that the percentage of remaining variation that is explained ranges from 2–3% to as high as 50%.

Although we have demonstrated that ignoring testlet effects, when they are there, yields problematic results, we still needed to see how often testlet effects show up in practice. To address this issue, we need to see whether testlet effects really existed, how large they were, and how hard we would have to look to find a data set that had them. To portend our findings:

i. They really exist.
ii. Sometimes their effect is very large, well over twice the variance of the abilities.
iii. You don't have to look hard to find them. In fact, in most data sets we observe with testlets, there are at least small to moderate testlet effects.

9.4. An application of the Bayesian 3-PL testlet model to GRE data

To put a "bow" on this method, we applied the Bayesian 3-PL testlet method to two real data sets from well-known operational large-scale tests. The first data set consisted of 4,000 examinees who had taken the verbal section of one administration of the 1996 Scholastic Assessment Test (SAT V). The other data set is from 5,000 examinees who had taken a verbal section of the 1997 Graduate Record Examination (GRE V). The structure of each of the two tests is given in Table 9.3.

Table 9.3. *The structure of the SAT V and GRE V data analyzed*

Test	Testlet	Number of items	σ_γ^2
SAT V	Independent	38	N/A
	1	12	0.13
	2	6	0.35
	3	10	0.33
	4	12	0.11
	Total	78	
GRE V	Independent	54	N/A
	1	7	0.61
	2	4	0.96
	3	7	0.60
	4	4	0.63
	Total	76	

We note that the SAT V section consisted of 78 items, 38 of which were independent items, whereas the other 40 were nested in testlets of sizes 12, 6, 10, and 12, respectively. For each of the SAT V testlets, the amount of estimated testlet variance is quite small, no larger than 0.35 (small compared to the variance of 1.0 for abilities). This suggests that either much of the prework done by the SAT V writers, who try to write items with good conditional independence properties, was effective, or possibly the items here are just not that locally dependent. In either case, more detailed results (not shown here) demonstrated that the differences between fitting the MCMC_d model (the "technically correct" model) and the $\text{MCMC}\gamma$, MCMC, and MML models did not make a significant difference.

In contrast to the SAT V analysis, the GRE V section was composed of 76 items, 54 of which were independent and the remaining 22 were nested within testlets of Sizes 7, 4, 7, and 4, respectively. Here, we note a much larger set of testlet variances, each of which is of a similar, albeit slightly smaller, order of magnitude than the variance of the abilities (equal to 1). Hence, we would expect to see more significant differences between non–testlet-based methods (MML and MCMC) and testlet-based methods for the GRE V data in comparison to the SAT V data.

In fact, this is what we found, and more. While, as expected, the parameter estimates for the GRE V data varied more between the MML and MCMC_d methods, we noticed another finding of interest. The differences between parameter estimates were significantly greater for the item discrimination parameters than the item difficulty or item guessing parameters. The reason for this is actually quite important and expected when looked at in the right way. It is well known that for the 3-PL testlet model the Fisher information (negative inverse of the asymptotic variance) for θ_i is given by

$$I(\theta_i) = a_j^2 \left(\frac{\exp(t_{ij})}{(1 + \exp(t_{ij}))} \right)^2 \frac{1 - c_j}{c_j + \exp(t_{ij})}, \tag{9.3}$$

where t_{ij} is the latent linear predictor of score as before. We note from (9.3) that the amount of information about the person-level ability is increasing in a_j, and one side effect of ignoring conditional independence is an increase in the assumed information, that is, a_j is larger. Hence, methods that ignore the testlet effects overestimate a_j, overstate the information, and hence yield asymptotic standard errors that are too small. This is what we did observe for the SAT V and GRE V data but (!) only for the testlet items and more so for the GRE V data where the testlet effects are larger. These insights while straightforward after the analysis were not obvious beforehand.

9.5. Summary

So, after extending the model in Chapter 8 to the 3-PL model, allowing for guessing to be explicitly incorporated, and by extending the model for testlet variances to be nonconstant across testlets, where are we? The model is now more realistic for binary items because it incorporates examinee guessing correctly as well as the possibility that not all testlets will have the same amount of testlet dependence. However, there are still two big holes left to fill. First, we need to be able to extend the methods to handle data that are not purely binary. Second, we need to be able to explain the "whys." Why do some testlets have more dependence than others? These are the topics of the next two chapters.

Questions

1. Write down the 3-PL model, and describe how it simplifies to the 2-PL model, the 1-PL model, and the Rasch model.
2. To demonstrate the weaker identifiability of the 3-PL model, construct two 3-PL item characteristic curves that have similar shapes but differing values of (a, b, c).
3. Explain how the testlet model with $\gamma_{id(j)} \sim N(0, \sigma^2_{d(j)})$ is related to the G-dimensional factor analysis model.
4. Describe a scheme in which the 3-PL model could be fit using data augmentation. Hint: introduce two latent augmented variables.

5. While the 3-PL model nests the 2-PL model, and hence by definition must have better in-sample fit, in an out-of-sample context it need not.
   i. Describe how would you assess the out-of-sample predictive power of the 3-PL and 2-PL models, and hence choose between them.
   ii. Describe how you could use Bayes factors to decide between the 3-PL and 2-PL models.
   iii. Explain what outcomes might be of interest to an educational researcher in assessing the predictive validity of one model versus another.
6. Write down a series of conditional distributions that would allow you to run a Bayesian version of the 3-PL TRT model described in this chapter. State which distributions can be sampled directly and which cannot. For those that cannot, describe how you would sample from their conditional distributions.

References

Birnbaum, A. (1968). Some latent trait models and their uses in inferring an examinee's ability. In F. M. Lord & M. R. Novick (Eds.), *Statistical theories of mental test scores* (pp. 395–479). Reading, MA: Addison-Wesley.

Bradlow, E. T. (2000). A comparison of computational methods for hierarchical models in customer survey questionnaire data. In R. S. Mariano, T. Schuermann, & M. Weeks (Eds.), *Simulation-based inference in econometrics: methods and applications* (pp. 307–327). London: Cambridge University Press.

Gelman, A., & Rubin, D. B. (1992). Inference from iterative simulation using multiple sequences. *Statistical Science, 7*, 457–472.

Wainer, H., Bradlow, E. T., & Du, Z. (2000). Testlet response theory: An analog for the 3-PL useful in testlet-based adaptive testing. In W. J. van der Linden, C. A. W. Glas (Eds.), *Computerized adaptive testing, theory and practice* (pp. 246–270). Boston, MA: Kluwer-Nijhoff.

10

A Bayesian testlet model for a mixture of binary and polytomous data

As the users of the scores from large-scale standardized tests have begun to demand more from tests than binary correct/incorrect answers can provide, test developers have turned increasingly to richer forms of assessment such as essays (Sireci & Zenisky, 2006), portfolios (Linn, Baker, & Dunbar, 1991), and even computer-displayed "cases" (Zenisky & Sireci, 2002). But these richer forms of assessment could not be scored sensibly as just correct/incorrect; it would be wasteful in the extreme to collapse all of the information in an examinee's essay to just a single bit of information. So, instead of a binary score, such complex tasks are typically scored polytomously. Such a scoring paradigm requires a more general psychometric scoring model to summarize components of tests, some of which might be scored in a binary way, others polytomously, and still others might be scored on a continuous scale.

Whether these forms of assessment, for the time that is spent constructing them, the amount it takes users to complete them, and the amount of time it takes to score them is more efficacious than just adding more multiple-choice items is an open empirical question and not one we shall address. However, this has been explored in part in Lukhele, Thissen, and Wainer (1994), and preliminary results suggest that from a purely information content perspective, multiple-choice items do pretty well. That is, if one wanted to just increase the reliability of the test such that the correlation between the observed score and the "true score" was maximized, multiple-choice items should play a significant part of tests. However, we would be remiss if we did not mention that increasing reliability or information content is not the reason that richer forms of assessment have become more prevalent. In fact, it is probably one of the least important reasons. We can easily think of two more important ones.

First, richer forms of assessment may have the potential to allow one to test *differing* aspects of ability. Many testing domains have abilities that are multifaceted, and essays, portfolios, and the like may allow the test developers to tap into all of those facets. Second, and maybe most important, one aspect of test development that is of premium importance is *predictive validity*; the degree to which the outcome of the test score predicts an exogenous outcome of interest. Thus, imagine one was interested in predicting how well a high-school student would write in college. It is not hard to argue that an essay component of an examination would be highly predictive of that. Imagine trying to predict how well a physician will examine patients once or if he or she is licensed. A test that actually has examinations of simulated patients, as part of them are much closer to the real event than simply asking a potential physician what he or she "would do" (Boulet et al., 1998). Hence, when one thinks of predictive validity, there appears to be a large role that nonbinary items might play, leading to outcome scores that are commonly ordinal in nature.

Simply put, ordinal data scores are here to stay, and, if anything, are increasing in their prevalence. Hence, the Bayesian testlet model for ordinal scores, described next, is not optional. The prevalence of such data in contemporary testing demands it.

10.1. Bayesian testlet model for binary and polytomous data

The Bayesian testlet model for mixed binary and polytomous data presented here is described fully in Wang, Bradlow, and Wainer (2002). The basic tenet of the model is that there are many testing situations in which tests have the following features:

i. binary scored items
ii. polytomously scored items, and
iii. testlets

all in the same test. Hence, the model presented here allows for tests that are composed of a mixture of binary items and polytomous items, which can be independent and/or nested within testlets. This is, of course, a completely general structure and is able to handle most practical testing situations that are faced today.

In particular, the model that we utilize for binary items is the 3-PL testlet model, described fully in equation (9.2) of the previous chapter, $p_{ij}(1) = P(y_{ij} = 1) = c_j + (1 - c_j)\,\text{logit}^{-1}(a_j(\theta_i - b_j - \gamma_{id(j)})$. We utilize this model here as it is the more general model, simplifies down to the 2-PL model when $c_j = 0$, simplifies down to standard models when $\gamma_{id(j)} = 0$, simplifies down

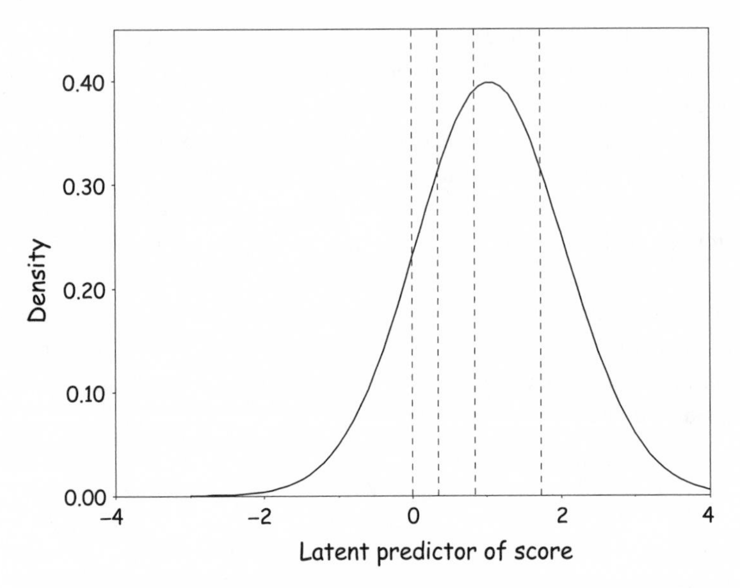

Figure 10.1. A graphical description of a polytomous test model. The dashed lines represent the latent cutoffs that determine the responses.

to the 1-PL model when $a_j = a$, $c_j = 0$, and hence gives us the full generality that we desire. In addition, the efficacy of the Bayesian 3-PL model has been demonstrated in Chapter 9, and hence supports its use in a mixed data format model.

There are a number of extant models available for the ordinal data part of the model. The one that we utilize here is the ordinal data model first introduced by Samejima (1969) as

$$p_{ij}(r) = P(y_{ij} = r) = P(y_{ij} = r|g,\ t_{ij}) = \Phi(g_r - t_{ij}) - \Phi(g_{r-1} - t_{ij}), \tag{10.1}$$

where $P(y_{ij} = r)$ is the probability that examinee i on item j receives score $r = 1, \ldots, R$; $g = (g_0, g_1, \ldots, g_{R-1}, g_R)$, the latent set of cutoffs that determines the observed response category (to be described more fully below), and $t_{ij} = a_j(\theta_i - b_j - \gamma_{id(j)})$, the latent linear predictor of score as described previously.

To aid in the intuition behind the polytomous model, consider the following interpretation. Imagine that when examinee i receives item j, the latent "ability" that is generated is t_{ij}, the latent linear predictor of score, which is assumed to be normally distributed with mean $a_j(\theta_i - b_j - \gamma_{id(j)})$ and variance 1 (to identify the model, the scale of the latent variable must be fixed at a constant, here arbitrarily, as is customary, set to 1). This situation is portrayed graphically in Figure 10.1.

Figure 10.1 depicts a response item with $R = 5$ possible responses, where the dotted lines indicate the estimated values of the cutoffs g. Depending on where t_{ij} falls with respect to latent cutoffs g_r, the respondent receives a score equal to the two cutoffs that their latent "ability" falls in between. That is, $P(y_{ij} = r) = P(t_{ij} < g_r) - P(t_{ij} < g_{r-1})$, which assuming normality and a variance of 1, is equal to $\Phi(g_r - t_{ij}) - \Phi(g_{r-1} - t_{ij})$, as given in (10.1). Note that to identify this model, besides setting the variance of $t_{ij} = 1$, we set $g_0 = -\infty$, $g_1 = 0$, and $g_R = \infty$. Hence, one only has to estimate $g_2, \ldots, g_{R-1}$.

Now, with the binary 3-PL testlet model in place, and the model in place for polytomous items, we obtain our mixed testlet model as simply the concatenation of these two models. The likelihood for the data matrix $Y = (y_{ij})$, where some items may be binary and some items may be polytomous is given by

$$p(Y|\Lambda_1) = \prod_{i=1}^{I} \left\{ \prod_{j \in \text{binary}} p_{ij}(1)^{y_{ij}} (1 - p_{ij}(1))^{1-y_{ij}} \right\} \left\{ \prod_{j \in \text{polytomous}} p_{ij}(r) \right\}, \tag{10.2}$$

where the first product is over all binary items and the second product is over all polytomous items.

To fully specify the integrated model, we need to utilize prior distributions for the parameters given in Λ_1. These priors are given by $a_j \sim N(\mu_a, \sigma_a^2)$, $b_j \sim N(\mu_b, \sigma_b^2)$, $\text{logit}(c_j) \sim N(\mu_c, \sigma_c^2)$, $\gamma_{id(j)} \sim N(0, \sigma_{d(j)}^2)$, and $\theta_i \sim N(0, 1)$, and new to this polytomous model $g_r \sim \text{uniform}(g_{r-1}, g_{r+1})$. To complete the model specification, we note that conjugate and slightly informative hyperpriors were utilized for all the parameters governing these priors.

As with the Bayesian 2-PL testlet model described in Chapter 8 and the Bayesian 3-PL testlet model described in Chapter 9, inferences under the model were performed by obtaining samples from the parameter's marginal posterior distributions using MCMC sampling. The details of the MCMC sampling are not provided here; however, the way in which the sampling proceeds is identical to before. We begin from an initial set of values for each of the parameters, and then we run Q independent chains from differing starting values. For each chain, and each of M iterations per chain, we sample from a subset of parameters in turn, conditioning on the current values of the other parameters and the data. After M' iterations, we deem the Markov chain to have converged, and then we utilize the remaining $Q \times (M - M')$ iterations for inference. The introduction of the polytomous form of the model, while requiring additional steps within each iteration, can be handled using the Metropolis–Hastings approach described in Chapter 9, and presents no significant obstacles. This again demonstrates the flexibility of the MCMC approach to handling extended models.

Table 10.1. *Simulation design for the Bayesian mixed testlet model*

		Factor A		
		2 categories	5 categories	10 categories
Factor B	6 testlets	0.0 (1)	0.5 (2)	1.0 (3)
	3 testlets	0.5 (4)	1.0 (5)	0.0 (6)
	2 testlets	1.0 (7)	0.0 (8)	0.5 (9)

We next describe a summary of an extensive set of simulations to test the efficacy of our MCMC approach to this model, and then we present an application of this Bayesian mixed testlet model to real data from the North Carolina Test of Computer Skills initially described in Rosa et al. (2001).

10.2. Simulation results for the Bayesian mixed testlet model

To demonstrate the Bayesian mixed testlet model, we ran simulations using three major design variables that would allow assessment of how the model would work under operational test conditions. The three major testlet design variables chosen were

Factor A: $R = (2, 5, \text{ or } 10)$, the number of response categories per testlet question;
Factor B: $D = (2, 3, \text{ or } 6)$, the number of testlets in the test, and
Factor C: Variance of the testlet effects (0, 0.5, or 1.0).

Factor A was selected to allow us to see whether the MCMC sampler would work when the number of response categories was small ($R = 2$, binary items), medium ($R = 5$), and large ($R = 10$). Note as well that running the model where $R = 2$ would allow us to reverify the previous chapter's results for binary items. Factor B was chosen to confirm the ability of the model when the number of items per testlet was small. Finally, as with all of the simulations that were run in all chapters, Factor C is utilized to see how well the model can capture the truth when the amount of local dependence varies.

Since running a full factorial design here, $27 = 3 \times 3 \times 3$ conditions would have been computationally time consuming, and since we were not particularly interested in all of the higher-order interactions, a latin-square design was used to determine the nine simulation conditions tested. These nine simulation conditions are given in Table 10.1, where the number in parentheses is our label for the simulation condition. Also note that Factor C's values are those that are

Table 10.2. *Correlation between true and estimated parameter values for the Bayesian mixed testlet model (R = number of response categories; D = number of testlets in the test)*

Simulation #	R	D	σ_γ^2	a	b	c	g_r	θ
1	2	6	0.0	0.89	0.99	0.64	NA	0.92
2	5	6	0.5	0.89	0.99	0.66	0.97	0.95
3	10	6	1.0	0.85	0.99	0.57	0.97	0.93
4	2	3	0.5	0.83	0.99	0.65	NA	0.91
5	5	3	1.0	0.89	0.99	0.60	0.98	0.90
6	10	3	0.0	0.95	0.99	0.62	0.99	0.98
7	2	2	1.0	0.87	0.98	0.50	NA	0.88
8	5	2	0.0	0.93	0.99	0.63	0.98	0.97
9	10	2	0.5	0.94	0.99	0.54	0.99	0.92

nested within the table. For instance, simulation condition 4 is the one in which there are two categories per question ($R = 2$), six testlets in the test ($D = 6$), and the variance of the testlet effects is 0.5.

Each of the nine simulation conditions was utilized to generate 5 data sets, yielding 45 data sets in total. Each data set contained simulated responses for $I = 1{,}000$ examinees of a test composed of $J = 30$ items. In each of the tests, the first 12 items were independent binary items, whereas the remaining 18 items were generated according to the design given in Table 10.1. As an example, in simulation condition (5), each of $I = 1{,}000$ simulees saw an examination in which the first 12 were independent binary items, the next 18 items were polytomous items with $R = 5$ response categories per item, which were nested into three testlets of Size 6, where $\sigma_\gamma^2 = 1.0$. The results presented below were derived from the average of the posterior means of five data sets conditions obtained by running an MCMC sampler in SCORIGHT 3.0 for $M = 3{,}000$ iterations utilizing the first $M' = 2{,}000$ as burn-in. Hence, each posterior mean is obtained from 1,000 draws.

As a very detailed summary of the simulation findings can be found elsewhere (Wang, Bradlow, & Wainer, 2002), we provide here only the most salient table from that paper. Table 10.2 presents the correlations between the true and estimated parameters from the Bayesian mixed testlet model.

These simulation results are extremely encouraging. First, the difficulty parameters b are well estimated by the model as are the latent cutoffs g_r. The remaining parameters from the model, a, c, and θ, are estimated exactly as in the previous simulation and as described in the extant literature. In addition, there was no pattern or degradation of the ability to recover parameters across conditions. With these simulations results now in place, we applied this model to real test data.

10.3. The application of the Bayesian mixed testlet model to the North Carolina Test of Computer Skills

The North Carolina Test of Computer Skills is an examination given to eighth graders that must be passed as a requirement for graduation from junior high school. It was developed to be part of a system to ensure basic computer proficiency among North Carolina school graduates. This examination had exactly the structure we were looking for. The first part of the exam is in a standard multiple-choice format while the second part is based on performance skills in four "testlet" areas: keyboarding, word processing/editing, database use, and spreadsheet use. The keyboarding portion includes three polytomous items scored on a four-point scale, whereas the remaining three testlet areas contained 6 to 10 items scored in either two or three categories.

Table 10.3. *Results from the Bayesian mixed testlet model fit to the North Carolina Test of Computer Skills*

	Number of polytomous items	Number of dichotomous items	Total	σ_γ^2
Keyboarding	3	0	3	0.03
Word processing	0	10	10	2.80
Database use	3	4	7	0.78
Spreadsheet use	1	5	6	2.58
Total	7	19	26	

The results reported here are the output from a single run of SCORIGHT 3.0 run on test data administered in the 1994–1995 school year generated by 266 participants. These were pilot data, and represented roughly 1/12th of the 3,099 participants who served in the pilot study. Each participant took the 26-item test designed as given in Table 10.3.

In the last column of Table 10.3 are the estimated values (posterior means) of σ_γ^2 for each of the four testlets. Noting as before that these values can be compared to $\sigma_\theta^2 = 1$ to get an order of magnitude, we see substantial testlet effects for word processing and spreadsheet use, and much lower ones for the other two testlets, being almost nonexistent for keyboarding. These results clearly demonstrate the need for the extension described in Chapter 9, in which the testlet variance was allowed to vary across the testlet.

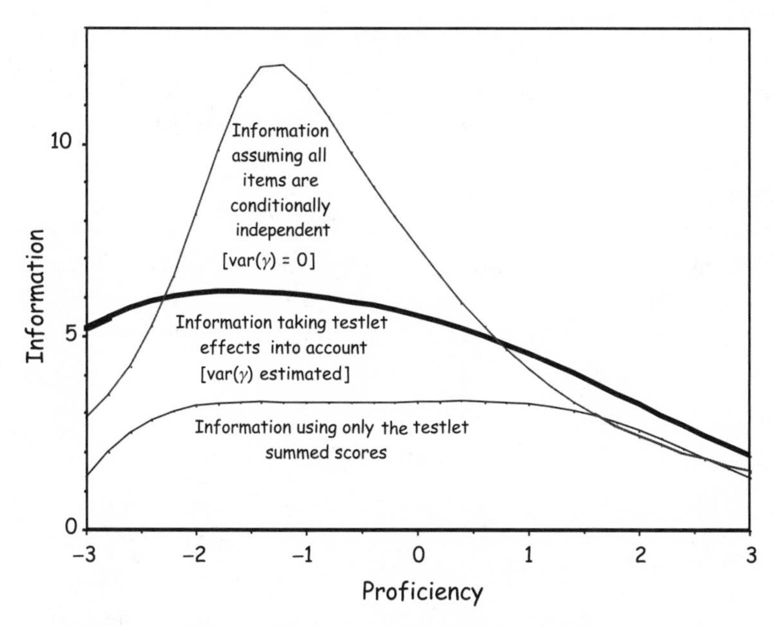

Figure 10.2. The information functions for the performance sections of the North Carolina Test of Computer Skills.

Another interesting finding from these results could be seen in a plot of the amount of information that is obtained by analyzing these data in three different ways.

The first way, which is depicted in the curve with the highest mode in Figure 10.2, is obtained by fitting a model assuming all 26 items are conditionally independent, that is, $\sigma_\gamma^2 = 0$. As Table 10.3 demonstrates, this is clearly erroneous, and not surprisingly leads to a gross overstatement of the amount of information available over a fairly wide range of the ability distribution. The second method, which corresponds to the middle curve in Figure 10.2, is obtained by using the Bayesian mixed testlet model described here. We note here that the information function is relatively flat and near its maximum over a wide range of values, which may be quite important (and good) if it corresponds to an area of the ability distribution in which the test developers want to estimate person's ability with highest precision. The last curve, the lowest in Figure 10.2, was obtained by collapsing each testlet into a single ordinal summed score (as described previously in Chapters 5 and 6). We see that while this leads to a simplified data structure that can be estimated using standard software, a fairly significant loss of information (around 50% in most cases) occurs in comparison with the Bayesian mixed testlet model.

10.4. A summary

The Bayesian mixed testlet model presented here represents a significant advance over the models described in Chapters 8 and 9, which allowed only for binary item responses. It is the extension to polytomous data that allows this model to have true operational impact on test practitioners. We turn next in Chapter 11 to the final "limitation" of the models described in Chapters 8 to 10, that is, answering the question of "why." In particular, in Chapter 11, we describe the use of covariates to explain the Bayesian 3-PL TRT model parameters, a topic of crucial contemporary scientific and practical importance.

Questions

1. An alternative to the polytomous model of Samejima utilized here is the graded response model where one fits a model to the $P(Y_{ij} \leq r)$ as opposed to $P(Y_{ij} = r)$. Compare and contrast these two models in general, and the role that cutoffs play in each of the two models.
2. Mentioned in this chapter are the tradeoffs between including items that are polytomous that provide richer data (although take more time) than do additional binary items. Describe in detail how would you determine which of the two is more "appropriate." In determining "appropriate," provide an answer in terms of parameter estimation, predictive validity, and practical concerns (test writing and scoring).
3. Write down how a data augmentation MCMC approach can be used for a polytomous test model to draw the parameters of the latent linear predictor of score t_{ij}.
4. Regarding Question (3), describe the tradeoffs between doing computation using data augmentation versus a Metropolis–Hastings approach.
5. Derive the elements of Fisher's information matrix for the Bayesian 3-PL mixed TRT model.
6. The information function in Figure 10.2 is relatively flat over the entire range of proficiency. Explain
   i. Why this might be a good thing?
   ii. When it might be more efficient to have a more peaked function?
   iii. How the "most desired" information function might differ for a pass/fail examination as compared to one in which a continuous score is given?

References

Boulet, J. R., Ben David, M. F., Ziv, A., Burdick, W. P., Curtis, M., Peitzman, S., & Gary, N. C. (1998). Using standardized patients to assess the interpersonal skills of physicians. *Academic Medicine, 73*, S94–S96.

Linn, R. E., Baker, E. L., & Dunbar, S. B. (1991). Complex, performance-based assessment: Expectations and validation criteria. *Educational Researcher, 20*(8), 15–21.

Lukhele, R., Thissen, D., & Wainer, H. (1994). On the relative value of multiple-choice, constructed response, and examinee-selected items on two achievement tests. *Journal of Educational Measurement, 31*, 234–250.

Rosa, K., Nelson, L., Swygert, K., & Thissen, D. (2001). Item response theory applied to combinations of multiple-choice and constructed-response items – Scale scores for patterns of summed scores. In D. Thissen & H. Wainer (Eds.), *Test scoring* (chap. 7, pp. 253–292). Hillsdale, NJ: Erlbaum.

Samejima, F. (1969). *Estimation of latent ability using a response pattern of graded scores* (Psychometrika Monographs Whole No. 17). Richmond, VA: Psychometric Society.

Sireci, S. G., & Zenisky, A. L. (2006). Innovative item formats in computer-based testing: In pursuit of improved construct representation. In S. M. Downing & T. M. Haladyna (Eds.), *Handbook of test development* (pp. 329–348). Mahwah, NJ: Erlbaum.

Wang, X., Bradlow, E. T., & Wainer, H. (2002). A general Bayesian model for testlets: Theory and applications. *Applied Psychological Measurement, 26*(1), 190–1128. Also listed as ETS GRE Technical Report 98-01.

Wang, X., Bradlow, E. T., & Wainer, H. (2004). A user's guide for SCORIGHT (version 3.0): A computer program built for scoring test built of testlets including a module for covariate analysis (ETS Research Report RR 04-49). Princeton, NJ: Educational Testing Service.

Zenisky, A. L., & Sireci, S. G. (2002). Technological innovations in large-scale assessment. *Applied Measurement in Education, 15*, 337–362.

11

A Bayesian testlet model with covariates

Chapters 7 to 10 provide a thorough introduction to a family of Bayesian-based testlet response models that will allow for the appropriate scoring of a broad range of tests. Yet, there remains one major theoretical hill to climb, as well as a minor technical one. In particular, while the model described in Chapter 10 allows us to model data composed of binary items, polytomous items, nested inside or outside of testlets, it does not help us explain "why" we observe what we do. This is our mission here.

Before we explicitly describe our approach, we note that for many practical situations, understanding the "why" may or may not be important. For the "is not important side," we can state that many tests are simply used as measurement devices, and what is important is that the device reliably measures the ability and item parameters as well as possible. A Bayesian mixed TRT model allows one to get "honest" estimates of examinee ability that integrates both multiple-choice and essay-like formats, and assumes conditional independence after incorporating the testlet structure. One does not need a theory or method of understanding "why" to answer this question. Pure measurement does not require a "why."

On the other hand, it is easy to imagine many settings where being able to understand "why" from a psychometric model is of primary interest. Three of these are

1. Imagine a situation in which it is important to understand whether certain populations or subgroups of individuals (as a whole) perform worse on the examination. This could be called DTF (differential test functioning) as compared to the normal acronym DIF (differential item functioning). Holland and Wainer (1993) provide many ways to look at this same question but only for single items. By bringing covariates into the model, for example using dummy variables to describe subgroup membership,

we can see whether there is a shift in abilities, θ_i, as a function of group membership.

2. Test designers can do their jobs better if they understand why certain items they construct are harder than others. For example, are there characteristics of the items themselves (number of words, topic area, item order) that would help one to explain some of the variation in difficulty, b_j?
3. As mentioned throughout Chapters 8 to 10, our extended model allows us to understand whether violations from conditional independence could be predicted from features of the testlets themselves, that is, can features of the testlets help predict $\sigma^2_{d(j)}$? What testlet structures have a propensity to induce greater dependence? This is an important topic for test designers.

We next describe our approach that could answer these questions and more. The three described here are meant only to be provocative and hopefully awakening about the power of the addition of covariates at many levels to the model. We should note that all of these inclusions are available in the current version of SCORIGHT.

11.1. An extended Bayesian TRT model

In many Bayesian statistical models, answering "why" questions is implemented by modeling parameters as a function of covariates. Specifically, since parameters in Bayesian models have prior distributions, covariates can be used to predict the mean of the prior of the distribution. Hence, if one were to observe a positive slope coefficient for a given covariate, it would indicate a higher value, on average, for that parameter, and vice versa across a heterogeneous population. It is in this manner that covariates via parameter distribution means are typically brought into Bayesian models in a natural way.

Before describing the full model, let us discuss the "technical hill" mentioned earlier that impacts the way we answer the "why" question. A limitation of the models presented in Chapters 8 to 10 is the assumption of independent prior distributions for the item parameters a_j, b_j, and c_j. This relates to the "weak identifiability" of the 3-PL model mentioned initially in Chapter 9, in which the estimates of the true latent values of the item parameters are correlated. For example, we might expect an item that has high difficulty to have high guessing. Or, we might expect items that are highly discriminating to be more difficult. Hence, rather than modeling each of the item parameters independently, in this chapter we allow for the item parameters' prior distributions to be correlated. This is continued throughout the remaining chapters of the book as well. The details are described next, and then, with this in place, we describe the insertion/utilization of covariates into the model.

The likelihood for observed responses y_{ij} that we utilize here is the same as that described in Chapter 10. For binary items, we utilize the 3-PL testlet model given by

$$p_{ij}(1) = P(y_{ij} = 1) = c_j + (1 - c_j)\text{logit}^{-1}(t_{ij}) \tag{11.1}$$

and for polytomous items, the testlet version of the Samejima (1969) ordered response model

$$p_{ij}(r) = P(y_{ij} = r) = P(y_{ij} = r \mid g, t_{ij}) = \Phi(g_r - t_{ij}) - \Phi(g_{r-1} - t_{ij}), \tag{11.2}$$

with $t_{ij} = a_j(\theta_i - b_j - \gamma_{id(j)})$. What differs significantly from previous chapters are the priors on the distributions for the parameters of (11.1) and (11.2).

Let $\lambda_{j,\text{3-PL}} = (\log(a_j), b_j, \text{logit}(c_j))$ denote the item parameter vector, and after transformation, for item j, a 3-PL binary item. We transform a_j using the log transformation and c_j, as in Chapters 9 and 10, using $\text{logit}(c_j) = \ln(c_j/(1 - c_j))$ so that both random variables after transformation are defined on the entire real line. This is important as our prior model for item j's parameters is given by

$$\lambda_{j,\text{3-PL}} \sim \text{MVN}(\mu_{j,\text{3-PL}}, \Sigma_{\text{3-PL}}), \tag{11.3}$$

where $\text{MVN}(v, w)$ stands for a multivariate normal random variable, with mean vector v and covariance matrix w. After transformation, it is reasonable to assume that $\log(a_j)$ and $\text{logit}(c_j)$ may be normally distributed; however, prior to transformation this would be unrealistic as a_j is usually positive and of course $0 \leq c_j \leq 1$.

This now allows for the two major changes described in this chapter. First, $\Sigma_{\text{3-PL}}$, as described above, allows the model to incorporate the correlation among the item parameters that is likely to exist. The simplified models in Chapters 8 to 10 are equivalent to assuming $\Sigma_{\text{3-PL}}$ is a diagonal 3×3 matrix. Whether the off-diagonal elements are zero now becomes an empirical question, which this model can address. Second, the way in which covariates are brought into the model is by specifying

$$\mu_{j,\text{3-PL}} = X'_{j,\text{3-PL}}\beta_{\text{3-PL}}, \tag{11.4}$$

where $X'_{j,\text{3-PL}}$ is a vector of covariates describing the j-th item that is a 3-PL item and $\beta_{\text{3-PL}}$ a set of covariate slopes for $\log(a_j)$, b_j, and $\text{logit}(c_j)$. Hence, $\beta_{\text{3-PL}}$ informs us as to the "why" certain 3-PL items may be found more or less difficult (higher b_j), more or less discriminating (higher a_j), or have a higher proportion of guessing (higher c_j).

We also allow a similar hierarchical prior structure for the polytomous items. The reason we need and incorporate a distinct structure here is that for the polytomous items there is no "guessing" parameter, and hence both the covariates that describe the parameters, and more importantly, the covariances among them may change. Thus, for the polytomous items, we model $\lambda_{j,\text{poly}} = (\log(a_j), b_j)$ as coming from

$$\lambda_{j,\text{poly}} \sim \text{MVN}(\mu_{j,\text{poly}}, \Sigma_{\text{poly}}), \tag{11.5}$$

where $\mu_{j,\text{poly}}$ is modeled as

$$\mu_{j,\text{poly}} = X'_{j,\text{poly}}\beta_{\text{poly}}, \tag{11.6}$$

and Σ_{poly} is a 2×2 covariance matrix. This model is general and allows different covariates to be available for modeling each parameter; we allow for different slope effects for each parameter, and allow different covariance matrices for each parameter set. To complete the prior model specification, we next describe the models, including covariates, for the abilities and testlet variances.

To allow the model to explain differences in abilities, θ_i, across subgroups, we utilize

$$\theta_i \sim N(X'_{i,\theta}\beta_\theta, 1). \tag{11.7}$$

The variance of this distribution is set to 1 to identify the model. In addition, there is no intercept in $X'_{i,\theta}$ and all covariates need to be mean-centered so that there will not be a shift identification problem with b_j and $\gamma_{id(j)}$, that is, you could add a constant to every θ_i in the model and subtract the same constant from b_j or $\gamma_{id(j)}$ and the likelihood would remain unchanged.

Finally, we introduce how covariates are brought into the testlet part of the model, which is the primary contribution of this chapter. For this part of the model, we utilize

$$\log\left(\sigma^2_{d(j)}\right) \sim N\left(X'_{\gamma d(j)}\beta_\gamma, \tau^2\right), \tag{11.8}$$

a log-normal prior for the testlet variances. Covariates that may be of use as part of $X'_{\gamma d(j)}$ might be the number of words in the testlet, the subject area of the testlet, the number of items in the testlet, or the location of the testlet in the overall test.

Unlike the models described in Chapters 7 to 10, the model described here, now implemented in SCORIGHT 3.0, is completely new and hence has yet to be utilized in a published study.

11.2. Example 1: USMLE data

The data we use to illustrate the TRT model with covariates were drawn from a multiple-choice component of Step 3 of the United States Medical Licensing Examination (USMLE™), the third in a sequence of tests required for medical licensure. Step 3 is designed to assess whether a physician possesses the qualities deemed essential to assume responsibility for providing unsupervised general medical care. Of primary concern is whether the physician has the requisite skills and can apply medical knowledge and understanding of clinical science in independent practice. Emphasis is placed on patient care and management in ambulatory settings.

To reduce computation time and to focus on items that are particularly relevant to our interest, 54 binary items were sampled from the multiple-choice portion of a single 282-item form. Eleven of these items were independent items and forty-three were items that were nested within 19 testlets. Fourteen of these testlets were composed of two items each; the remaining five testlets were composed of three items each. A typical testlet consists of a vignette that describes a medical situation and then one or more items that probes aspects of that situation. The data contained a complete set of binary responses for 905 examinees.

To understand the characteristics of the items and examinees, we incorporated several covariates into the analysis. To understand more about the item parameters, we include three covariates: vignette word count, stem word count, and options word count. *Vignette word count* is the number of words contained in the core reading passage itself, while *stem word count* and *options word count* are the numbers in the actual question and in the various choices available as responses, respectively. To understand the testlets, better we include the same three covariates, which are also collected for each testlet. To understand the proficiency of examinees, we consider five factors: gender, ethnicity, native English speaker or not, median item response time, and whether or not an examinee belongs to LCME, a group that certifies medical school programs. By considering all the covariates of examinees, we have 729 examinees with complete records among the 905. Therefore, 729 examinees' responses to the 54 items were used in the analysis. Among these 729 examinees, 56% were male and 44% were female; 56% had the LCME certification and 44% did not; 58%

were native English speakers and 42% were not; 30% were Asian, 5% were Black, 7% were Hispanic, 50% were White, and 8% were classified as "other."

11.2.1. Modeling the data

All the USMLE items were analyzed by utilizing the 2-PL logistic model, since the guessing-level parameter was deemed unnecessary, as it did not significantly improve model fit. Because there are 19 testlets in the test, (11.1) becomes

$$p_{ij}(1) = P(y_{ij} = 1) = \text{logit}^{-1}\left(a_j\left(\theta_i - b_j - \gamma_{id(j)}\right)\right), \tag{11.9}$$

which is a special case with $c_j = 0$. We used the three covariates described above for both item discrimination parameter a_j and item difficulty parameter b_j.

For each item j, let $x_{j,1}$ represent vignette word count, $x_{j,2}$ represent stem word count, and $x_{j,3}$ represent options word count. Then in (11.4), $X'_{j,\text{3-PL}}$ becomes

$$X'_{j,\text{2-PL}} = \begin{pmatrix} 1 & x_{j,1} & x_{j,2} & x_{j,3} & 0 & 0 & 0 & 0 \\ 0 & 0 & 0 & 0 & 1 & x_{j,1} & x_{j,2} & x_{j,3} \end{pmatrix}.$$

Notice that the entry in the first column and the first row of $X'_{j,\text{2-PL}}$ corresponds to the intercept for the prior mean of $\log(a_j)$, whereas the second to the fourth columns of the first row correspond to the three covariates: vignette word count, stem word count, and options word count. The rest of the entries of the first column are set to 0. The first four entries in the second row of $X'_{j,\text{2-PL}}$ are set to 0, while the fifth to the eighth entries of the second row correspond to the intercept and the three covariates for the prior mean of item parameter b_j. For both $\log(a_j)$ and b_j, $(x_{j,1}\, x_{j,2}\, x_{j,3})$ correspond to vignette word count, stem word count, and options word count (mean centered) at the item level. Hence,

$$E\begin{pmatrix} \log(a_j) \\ b_j \end{pmatrix} = X'_{j,\text{2-PL}}\beta_{\text{2-PL}} = \begin{pmatrix} 1 & x_{j,1} & x_{j,2} & x_{j,3} & 0 & 0 & 0 & 0 \\ 0 & 0 & 0 & 0 & 1 & x_{j,1} & x_{j,2} & x_{j,3} \end{pmatrix} \begin{pmatrix} \beta_a \\ \beta_{a,1} \\ \beta_{a,2} \\ \beta_{a,3} \\ \beta_b \\ \beta_{b,1} \\ \beta_{b,2} \\ \beta_{b,3} \end{pmatrix}.$$

For each examinee, by considering the five person-level factors described earlier, we have eight covariates in $X'_{i,\theta}$. They are

- $x_{i,1} = 1,$ if an examine is male; 0 if an examinee is female;
- $x_{i,2} = 1,$ if an examineee belongs to LCME group; 0 if he or she does not;

- $x_{i,3} = 1$, if an examinee is a native English speaker, 0 if he or she is not;
- $x_{i,4} = 100$/median item response time (in seconds) so that higher response times are represented by slower speeds;
- $x_{i,5} = 1$, if an examinee is Asian; 0 otherwise;
- $x_{i,6} = 1$, if an examinee is Hispanic; 0 otherwise;
- $x_{i,7} = 1$, if an examinee is Black; 0 otherwise;
- $x_{i,8} = 1$, if an examinee is White; 0 otherwise.

Then $X'_{i,\theta}$ in (11.7) is

$$X'_{i,\theta} = (w_{i,1}, w_{i,2}, w_{i,3}, w_{i,4}, w_{i,5}, w_{i,6}, w_{i,7}, w_{i,8}),$$

where $w_{i,j}$'s are mean centered $x_{i,j}$'s as before. Hence,

$$\begin{aligned} E(\theta_i) &= X'_{i,\theta}\beta_\theta \\ &= w_{i,1}\beta_{\theta,1} + w_{i,2}\beta_{\theta,2} + w_{i,3}\beta_{\theta,3} + w_{i,4}\beta_{\theta,4} + w_{i,5}\beta_{\theta,5} \\ &\quad + w_{i,6}\beta_{\theta,6} + w_{i,7}\beta_{\theta,7} + w_{i,8}\beta_{\theta,8}. \end{aligned}$$

We use the same three variables as covariates for the testlet effect as for the item parameters. However, the difference is that the information of word count based on vignette, stem, and options are collected at the testlet level instead of the item level as before. To distinguish these from the covariates at the item level, we denote the three covariates for the testlet effect as $z_{d(j),1}$ for vignette word count, $z_{d(j),2}$ for stem word count, and $z_{d(j),3}$ for options word count. Therefore, $X'_{\gamma d(j)}$ in (11.8) is given as

$$X'_{\gamma d(j)} = \left(1, z_{d(j),1}, z_{d(j),2}, z_{d(j),3}\right).$$

Hence,

$$E\left(\log\left(\sigma^2_{d(j)}\right)\right) = X'_{\gamma d(j)}\beta_\gamma = \beta_{\gamma,0} + z_{d(j),1}\beta_{\gamma,1} + z_{d(j),2}\beta_{\gamma,2} + z_{d(j),3}\beta_{\gamma,3}.$$

To summarize the parameters in the model, we have Λ_1, corresponding to the 2-PL TRT model as given in Chapters 8 to 10:

$$\Lambda_1 = \left\{\theta_i, a_j, b_j, \gamma_{id(j)}, \sigma^2_{d(j)}\right\},$$

and also several new parameters, Λ_2, that correspond to the covariate slopes and covariance matrices between the item parameters,

$$\Lambda_2 = \{\beta_\theta, \beta_{\text{2-PL}}, \Sigma_{\text{2-PL}}, \beta_\gamma, \tau\}.$$

To apply the Bayesian paradigm, we denote the joint posterior distribution of the 2-PL logistic model of (11.9) and the parameter Λ_1 and Λ_2 by $P(\Lambda_1, \Lambda_2 \mid \mathbf{y})$. Then by Bayes theorem, we have

$$P(\Lambda_1, \Lambda_2 \mid \mathbf{y}) \propto \mathcal{L}(\mathbf{y} \mid \Lambda_1, \Lambda_2)\, p(\Lambda_1 \mid \Lambda_2)\, p(\Lambda_2).$$

Here $\mathcal{L}(y \mid \Lambda_1, \Lambda_2)$ is the likelihood of the data given the model and the model's parameters. The prior distributions for elements of Λ_1 given Λ_2 are described in the previous section. To obtain the posterior distribution, it remains only to specify a suitable prior distribution (hyperpriors) for the parameters of Λ_2.

Here $\beta_{\text{2-PL}}$ and $\Sigma_{\text{2-PL}}$ are "hyperparameters" that determine the priors and are chosen as follows: $\beta_{\text{2-PL}} \sim N(0, V)$, where $|V|^{-1}$ is set to 0 to give a noninformative prior; $\Sigma_{\text{2-PL}}$ is set as an inverse-Wishart prior where the two parameters of inverse-Wishart distribution are 2 (minimally proper) and $\mathbf{M}^{-1}$, where $\mathbf{M} = \sigma^2\mathbf{I}$ with a small value of σ^2 to give an essentially noninformative prior and I is a 2×2 identity matrix. The prior distributions of β_θ and β_γ are assumed to be the same noninformative prior distributions as for $\beta_{\text{2-PL}}$. For the prior of τ^2, we chose an inverse-gamma distribution with both parameters close to 0, which also yields an essentially noninformative prior. We also assume independence among the prior distributions of $(\beta_{\text{2-PL}}, \Sigma_{\text{2-PL}})$, β_θ, β_γ, and τ^2. SCORIGHT then draws the posterior samples for each parameter on the basis of the following posterior distribution:

$$\begin{aligned}
&P(\Lambda_1, \Lambda_2, \mid y) \\
&\propto \prod_{i=1}^{I}\prod_{j=1}^{J}\left[\frac{\exp\left(e^{\log(a_j)}(\theta_i - b_j - \gamma_{id(j)})\right)}{1+\exp\left(e^{\log(a_j)}(\theta_i - b_j - \gamma_{id(j)})\right)}\right]^{y_{ij}}\left[\frac{1}{1+\exp\left(e^{\log(a_j)}(\theta_i - b_j - \gamma_{id(j)})\right)}\right]^{1-y_{ij}} \\
&\times \prod_{i=1}^{I}\exp\left(-\frac{1}{2}(\theta_i - X'_{i\theta}\beta_\theta)^2\right)\exp\left[-\frac{1}{8}(\beta'_\theta\beta_\theta)\right] \\
&\times \prod_{j=1}^{J}\left\{|\Sigma_{\text{2-PL}}|^{-\frac{1}{2}}\exp\left[-\frac{1}{2}(\lambda_{j,\text{2-PL}} - X'_{j,\text{2-PL}}\beta_{\text{2-PL}})'\Sigma^{-1}_{\text{2-PL}}(\lambda_{j,\text{2-PL}} - X'_{j,\text{2-PL}}\beta_{\text{2-PL}})\right]\right\} \\
&\times |\Sigma_{\text{2-PL}}|^{-\frac{5}{2}}\exp\left(-\frac{1}{200}\operatorname{tr}(\Sigma^{-1}_{\text{2-PL}})\right) \\
&\times \prod_{d(j)=1}^{K}\left[\prod_{i=1}^{I}\left(\frac{1}{\sigma_{d(j)}}\exp\left(-\frac{1}{2\sigma^2_{d(j)}}\gamma^2_{id(j)}\right)\right)\sigma^{-2}_{d(j)}\right. \\
&\left.\times \exp\left(-\frac{1}{2\tau^2}\left[\log\left(\sigma^2_{d(j)}\right) - X'_{\gamma d(j)}\beta_\gamma\right]^2\right)\right]\tau^{-2\left(\frac{1}{4}+1\right)}\exp\left(-\frac{1}{2\tau^2}\right). \qquad (11.10)
\end{aligned}$$

The difference between using a Bayesian framework to answer the "whys" and running a postanalyses regression directly based on the posterior means of the θ_i's, a_j's, b_j's, and $\log(\sigma^2_{d(j)})$'s on the corresponding covariates is that the posterior samples, and hence estimates, are based on different posterior distributions. For example, without covariates, θ_i is drawn on the basis of the following conditional distribution:

$$\prod_{j=1}^{J}\left[\frac{\exp\left(e^{\log(a_j)}(\theta_i - b_j - \gamma_{id(j)})\right)}{1+\exp\left(e^{\log(a_j)}(\theta_i - b_j - \gamma_{id(j)})\right)}\right]^{y_{ij}}\left[\frac{1}{1+\exp\left(e^{\log(a_j)}(\theta_i - b_j - \gamma_{id(j)})\right)}\right]^{1-y_{ij}} \times \exp\left(-\frac{1}{2}\theta_i^2\right). \tag{11.11}$$

With covariates, it is drawn on the basis of

$$\prod_{j=1}^{J}\left[\frac{\exp\left(e^{\log(a_j)}\left(\theta_i - b_j - \gamma_{id(j)}\right)\right)}{1+\exp\left(e^{\log(a_j)}\left(\theta_i - b_j - \gamma_{id(j)}\right)\right)}\right]^{y_{ij}}\left[\frac{1}{1+\exp\left(e^{\log(a_j)}\left(\theta_i - b_j - \gamma_{id(j)}\right)\right)}\right]^{1-y_{ij}} \times \exp\left(-\frac{1}{2}\left(\theta_i - X'_{i\theta}\beta_\theta\right)^2\right). \tag{11.12}$$

Conditional distribution (11.12) takes the uncertainty of the estimates into account, thus providing more realistic estimates of the corresponding slopes than does a direct regression using, say, a posterior mean as a dependent variable. In addition, since the effects of covariates are partialled out directly, there can be more stable estimates of each result.

11.2.2. The results of the covariate augmented analysis

Three Markov chain Monte Carlo chains (MCMC) were run, each with random initial values. Each chain had 20,000 draws, and the first 10,000 draws were discarded for burn-in. Among the last 10,000 draws for each chain, 1 out of every 20 were recorded to avoid the high autocorrelations among the posterior samples (see Chapter 15 for a further discussion). The Bayesian posterior estimates for each parameter were then based on those 1,500 draws combined from all three samples.

From SCORIGHT's diagnostics, it was noted that all six posterior samples had converged properly after the 10,000 burn-in draws. In general, without covariates, the MCMC sample will converge much faster. Typically, a data set of this size, without covariates, will converge after about 5,000 iterations. In this case, the extended parameter vector causes the Markov chain to move more slowly through the parameter space.

For the estimated coefficients of the covariates of the item discrimination and item difficulty parameters, the posterior mean and the posterior standard deviations are given in Table 11.1. For the item discrimination parameters, the

Table 11.1. *Estimated coefficients of the three covariates for the item discrimination and item difficulty parameters*

	Variable	Posterior mean	Posterior standard deviation
Item discrimination	Intercept	−1.025	0.324
a	Vignette word count	−0.001	0.002
	Stem word count	−0.003	0.003
	Options word count	0.01	0.006
Item difficulty	Intercept	−4.13	1.58
b	Vignette word count	0.002	0.008
	Stem word count	−0.001	0.017
	Options word count	0.017	0.031

Table 11.2. *Estimated coefficients of the three covariates for testlet effect* $[\log(\sigma^2_{d(j)})]$

Variable	Posterior mean	Posterior standard deviation
Intercept	1.87	0.90
Vignette word count	0.000	0.005
Stem word count	−0.009	0.023
Options word count	−0.007	0.010

estimated slopes for the three covariates are all close to 0. The three plots in the first row of Figure 11.1 show the kernel density estimates for the coefficients of the three covariates based on their posterior samples. None of them shows strong evidence that they are different from 0; albeit there is some evidence that option word count is positively associated ($p \approx 0.05$) with the item discrimination mean. Thus, it suggests that items with more words tend to be slightly more discriminating. In summary, item discrimination, a, is not affected by vignette word count or stem word count, and is moderately so by options word count. For the item difficulty parameters, all null results are observed as shown in both the bottom of Table 11.1 and the bottom three panels in Figure 11.1.

For the testlet effect variances, we have also included the same three covariates as for the item parameters, but now the three are computed at the testlet level. According to the posterior means, the estimated coefficients are all close to 0. We are thus led to the same inference about the value of these covariates for testlets as we drew for item discrimination and item difficulty. Table 11.2 shows the posterior means and standard deviations for the estimated coefficients, and

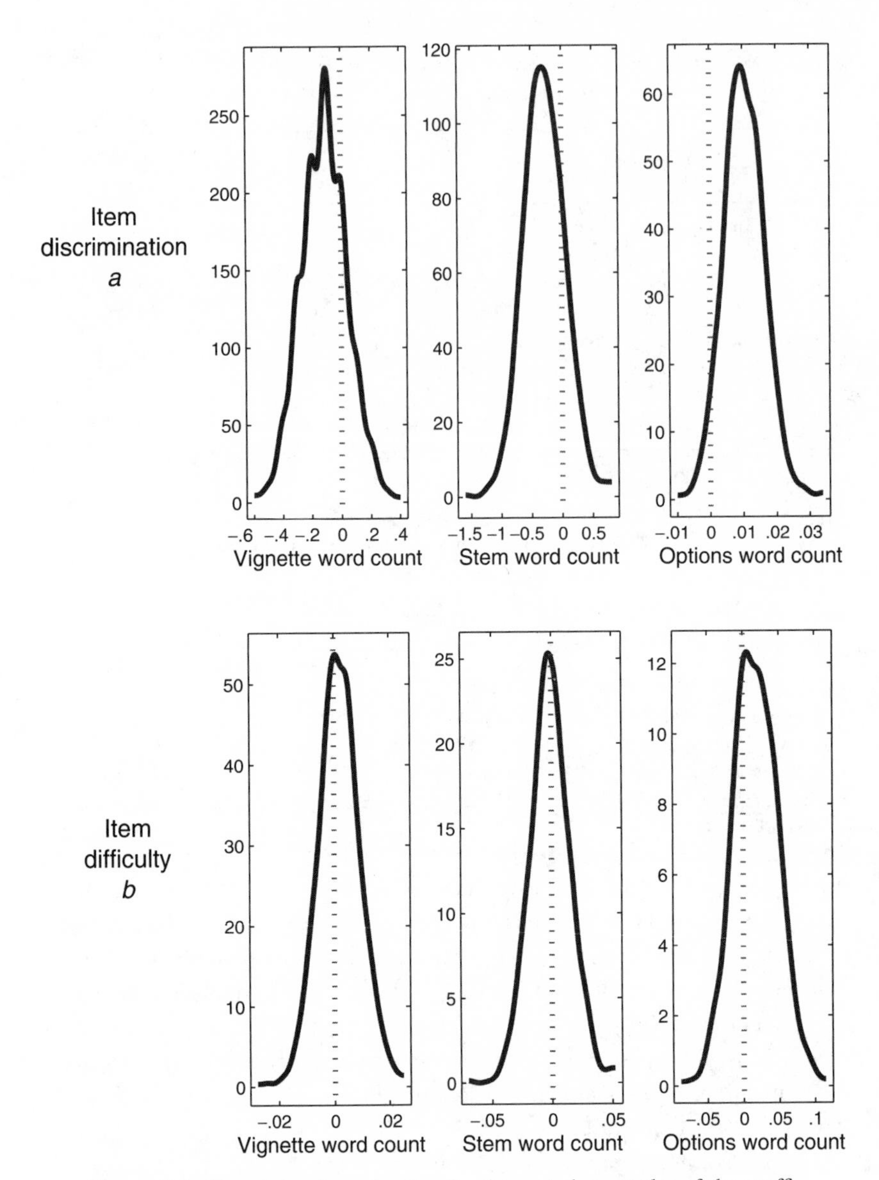

Figure 11.1. Kernel density estimation for the posterior samples of the coefficients for the three covariates for both item discrimination parameter a's and item difficulty parameter b's. A dashed vertical line is drawn at 0, indicating no effect.

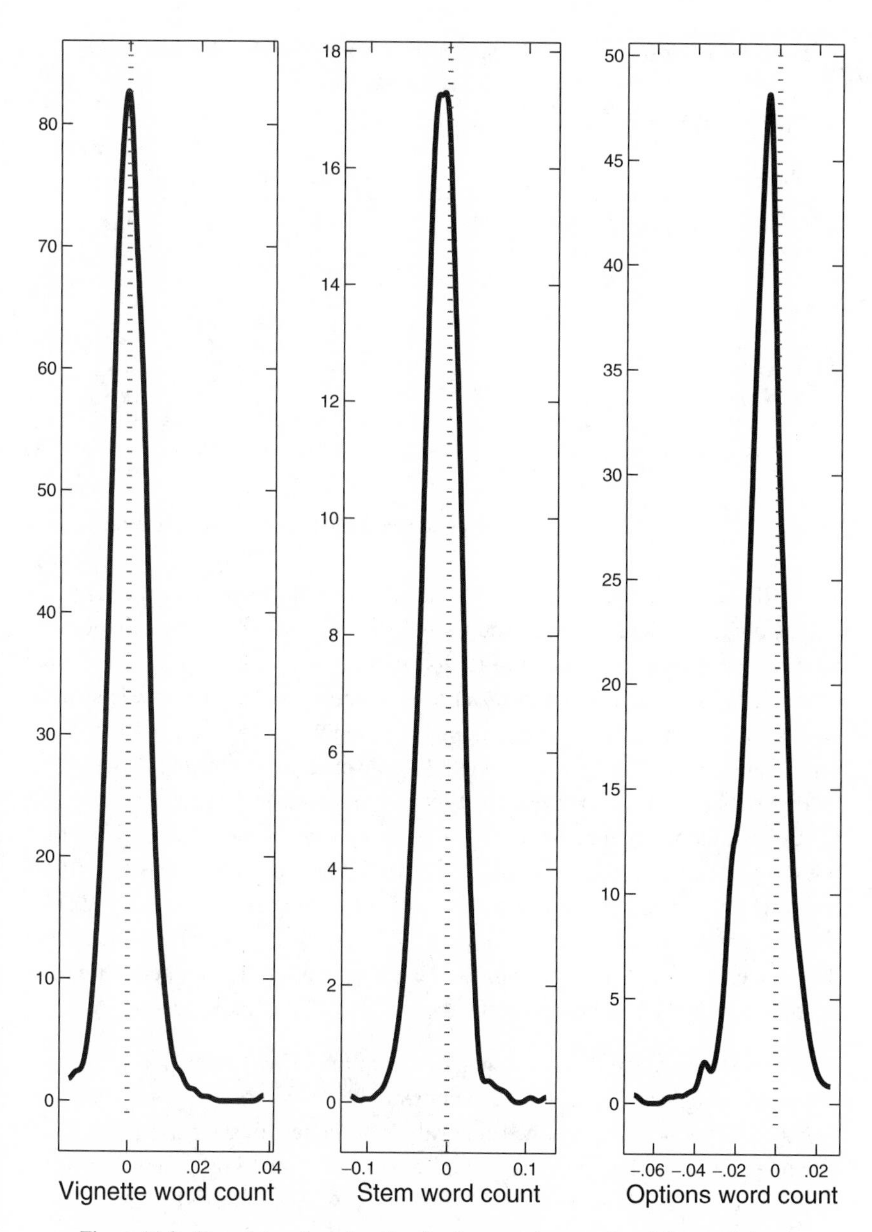

Figure 11.2. Kernel density estimation for the posterior samples of the coefficients of the three covariates for testlet effect [$\log(\sigma^2_{d(j)})$].

Table 11.3. *Estimated coefficients of the eight covariates for examinees' proficiency θ, $X_{i,\theta}$*

Variable		Posterior mean	Posterior standard deviation
100/median item response time	$\beta_{\theta,4}$	1.09	0.18
LCME group (yes = 1)	$\beta_{\theta,2}$	0.90	0.13
English native speaker (yes = 1)	$\beta_{\theta,3}$	0.43	0.14
Gender (male = 1)	$\beta_{\theta,1}$	0.06	0.11
White (yes = 1)	$\beta_{\theta,8}$	0.29	0.20
Asian (yes = 1)	$\beta_{\theta,5}$	−0.17	0.20
Hispanic (yes = 1)	$\beta_{\theta,6}$	−0.29	0.25
Black (yes = 1)	$\beta_{\theta,7}$	−0.60	0.27

Figure 11.2 shows their corresponding kernel density estimates based on their posterior samples.

For the examinee proficiency covariates, β_θ, the Bayesian estimates for the coefficients of gender, LCME group, English native speaker, and 100/median item response time are given in the first four rows of Table 11.3. The effect of ethnicity is divided into four covariates: Asian versus non-Asian, Hispanic versus non-Hispanic, Black versus non-Black, and White versus non White. The Bayesian estimates for their coefficients are given at the bottom in Table 11.3.

Figure 11.3 shows the kernel density estimates based on the posterior samples for the estimated coefficients of the first four covariates. On the basis of the posterior samples, we can calculate the 95% highest posterior density (HPD) intervals for each slope. For example, the 95% HPD interval for $\beta_{\theta,1}$ is (−0.14, 0.27). From the upper-left plot of Figure 11.3, it is clear that there is no gender difference. From the rest of the plots of Figure 11.3, all three other estimated coefficients showed significant differences from 0. They indicate that

- there is strong evidence that examinees from the LCME group have generally higher proficiency (0.90 units higher) than those who are not from that group, conditional on the same values of the rest of the parameters;
- native language plays an important role when evaluating proficiency. With the same values of other variables, examinees who are native English speakers have an average proficiency about 0.43 units higher than nonnative English speakers;
- the speed of response also plays an important role. The faster the response time is, the higher the estimated proficiency (we make no claims about the direction of the causal arrow).

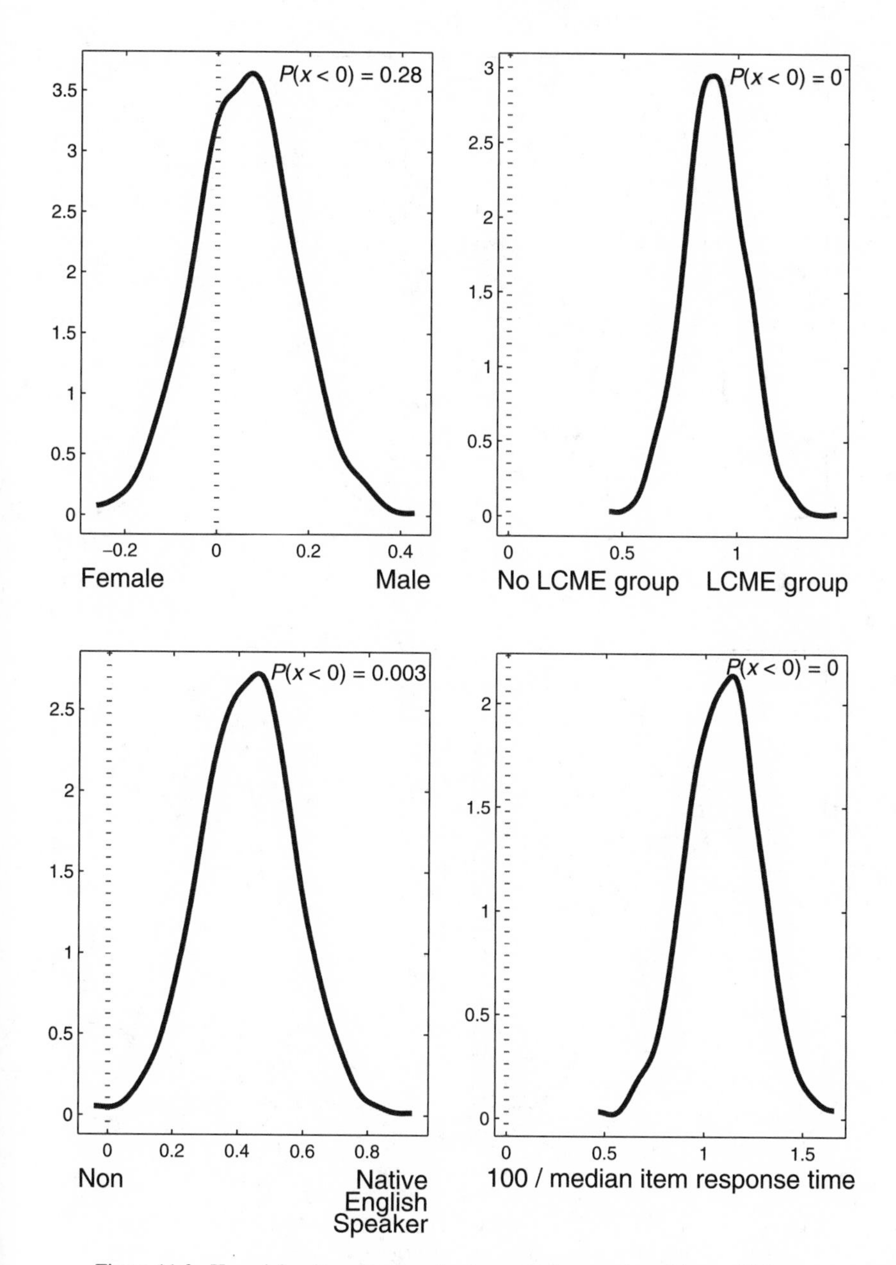

Figure 11.3. Kernel density estimation for the posterior samples of the coefficients of the first four covariates for examinees' proficiency θ.

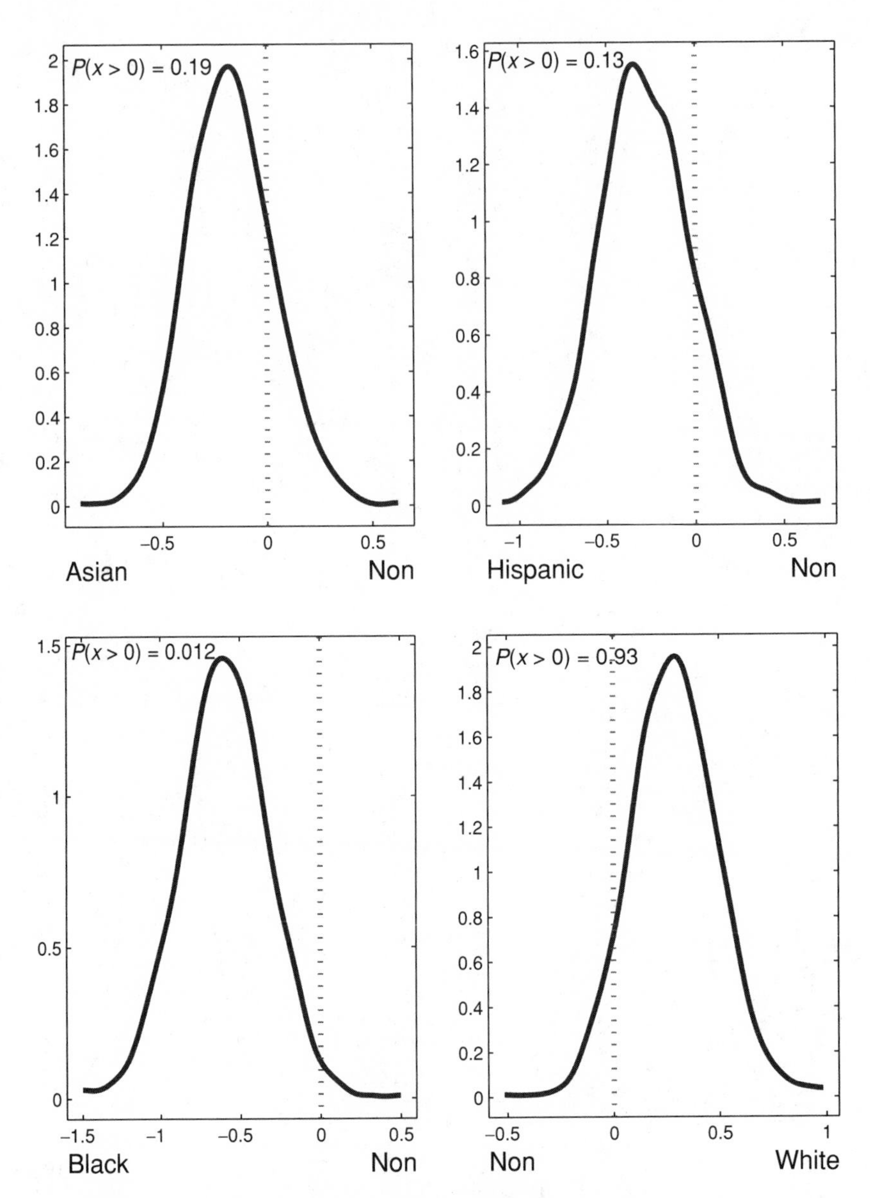

Figure 11.4. Kernel density estimation for the coefficients of the four group membership covariate effects for examinees' proficiency θ.

Table 11.4. *p-values for the difference of proficiency of six ethnic groups*

	p value
Asian–Black	0.90
Hispanic–Black	0.80
Asian–Hispanic	0.64
Asian–White	0.06
Hispanic–White	0.03
Black–White	0.004

Figure 11.4 shows the kernel density estimates for the four coefficients of the dummy variables of ethnicity. From these plots, it is easy to see the impact of a group having a particular characteristic or not (e.g., Asian vs. non-Asian). For instance, in this case, we can see no significant effect. However, in contrast, it is hard to see the difference between different ethnic groups (e.g., Asians vs. Whites) based solely on these plots; but since SCORIGHT saves all the posterior samples for each coefficient, we can compare the difference between any pair of ethnic groups directly. Figure 11.5 shows the plots of differences between six pairs of groups. They are Asian versus Hispanic, Asian versus Black, Asian versus White, Hispanic versus Black, Hispanic versus White, and Black versus White. The corresponding probabilities for the differences (one-sided) are given in Table 11.4. The results show significant differences between Black and White as well as Hispanic and White. There was a moderate difference between Asian and White examinees, with no other significant effects. Again, within the Bayesian framework, these inferences come for "free" along with parameter inferences for the a's, b's, and c's.

11.3. Example 2: Using testlet response theory to understand a survey of patients with breast cancer[1]

11.3.1. Introduction

Surveys are typically constructed with two related but different goals in mind. The first, and the easiest to accomplish, is the retrieval of facts from the surveyed

[1] Much of this analysis was done with the crucial help of Su Baldwin. In addition, the data were supplied by Bryce Reeve of the National Cancer Institute. We would like to express our gratitude to both of these generous scholars.

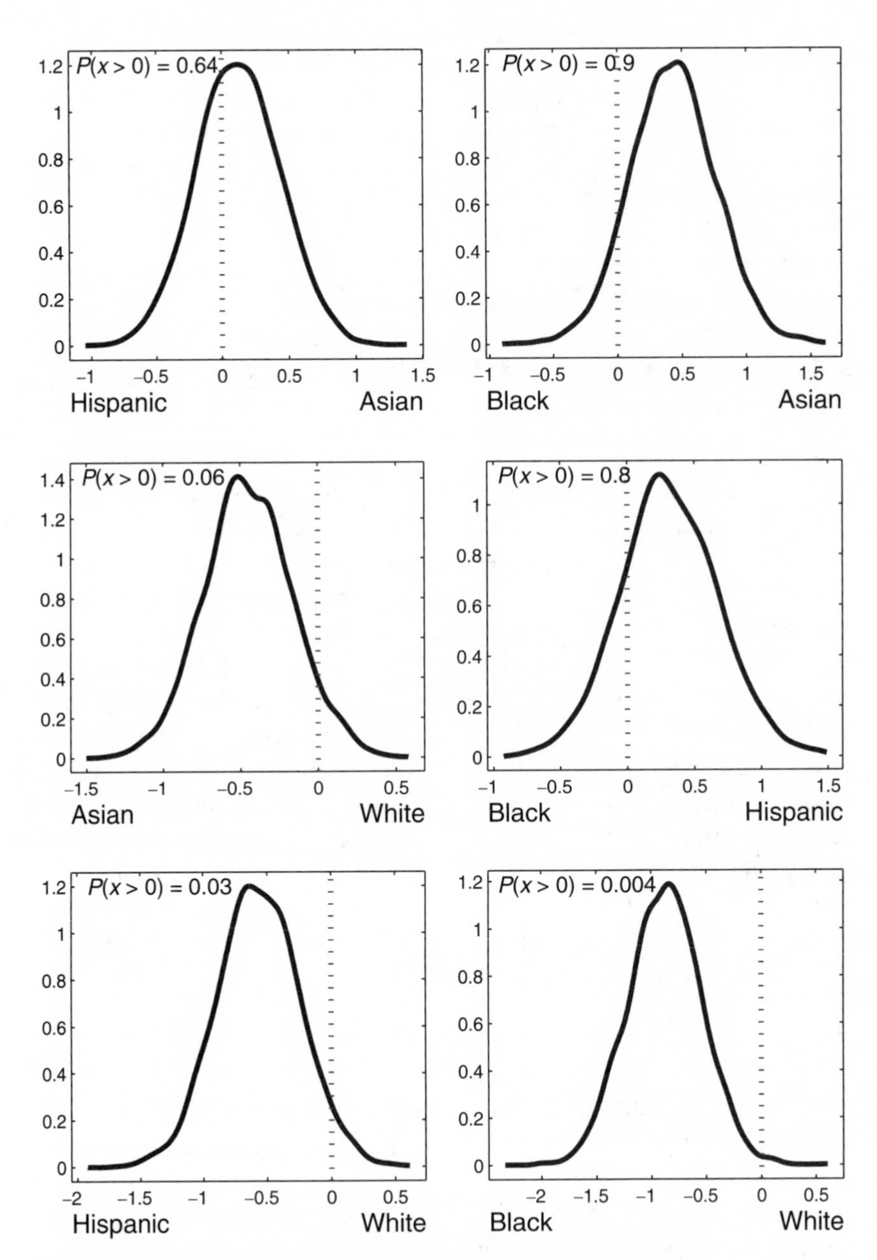

Figure 11.5. Proficiency differences between six pairs of ethnic groups.

population (e.g., "If the election were to be held tomorrow would you vote for A or B?"). The second goal is to retrieve information about a more vague concept, or one that may be conceived as a "higher order" construct or factor (Spearman, 1904), that cannot be directly addressed with a single, factual question. Instead, we try to tap into it with a sequence of questions that all (individually) shed a little light on the underlying variable of interest, but as a whole give us a fairly complete understanding of the construct of interest. For example, if we are interested in studying the level of health in a society, we might ask questions that reflect differing amounts of robustness, and/or tap into different aspects of health (physical, mental, short-term vs. enduring, etc.).

In addition to questions about the issue of interest, surveys also usually try to divide up the surveyed population using various demographic/background questions. Such information is included to help us understand any systematic variations in the responses to the questions asked as a function of those background variables (e.g., Do women tend to favor candidate A more than men do? Do Hispanics enjoy more robust health than Blacks?). These results can then be used for issues such as targeting underserved segments, or merely for descriptive purposes and reporting.

Last, and most relevant for this book, survey questions often are administered in testlets, groups of questions each of which focuses on a single theme (e.g., perceived satisfaction). As we have seen, this characteristic limits the range of measurement models that can be used to analyze such data to those that do not depend on the assumption of conditional local independence among all of the items within the survey.

In this example, we examine a 21-item survey composed of five testlets that was administered to 718 women being treated for breast cancer. Each item measured a patient's self-reported degree of perceived change on a 0 to 5-point Likert-type scale, ranging from "no change" to "very great change." One goal of the survey was to measure the extent to which these women's attitudes changed after diagnosis, and was based on the posttraumatic growth inventory developed by Tedeschi and Calhoun (1996). A list of the survey items is given in Table 11.5 and the five testlets were:

I. Relating to others
II. New possibilities
III. Personal strength
IV. Spiritual change
V. Appreciation of life

In addition to the 21 survey items, there were also background questions that asked about race, ethnicity, age, income, employment and marital status,

Table 11.5. *The survey items ordered by the testlet they were written to represent*

Testlet number	Item number	Text of item
I	6	Knowing that I can count on people in times of trouble
	7	A sense of closeness with others
	9	A willingness to express my emotion
	12	Having compassion for others
	15	Putting effort into my relationships
	18	I learned a great deal about how wonderful people are
	20	I accept needing others
II	19	I developed new interests
	21	I established a new path for my life
	13	I'm able to do better things with my life
	14	New opportunities are available which wouldn't have been otherwise
	2	I'm more likely to try to change things which need changing
III	4	A feeling of self-reliance
	8	Knowing I can handle difficulties
	10	Being able to accept the way things work out
	17	I discovered that I'm stronger than I thought I was
IV	5	A better understanding of spiritual matters
	16	I have a stronger religious faith
V	1	My priorities about what is important in life
	3	An appreciation for the value of my own life
	11	Appreciating each day

whether they were taking Tamoxifen (a drug commonly used to assist in the prevention and recurrence of breast cancer in women near or beyond menopause[2]), and how long had it been since they were diagnosed. The survey questions and the testlets they fell into are shown in Table 11.5.

This is a nonfactual survey in which each item is a manifest indicator of a latent trait of interest.

A complete analysis of the data from this survey would estimate each respondent's position on a latent trait underlying the question responses while dealing with the local dependence engendered by the survey construction. Then utilize the background information as covariates to help explain the results, that is, answer why. In the balance of this example, we describe both the method of analysis used and the results obtained. In addition, at various points in the

[2] A detailed description can be found at http://www.breastcancer.org/tre_sys_tamox_idx.html.

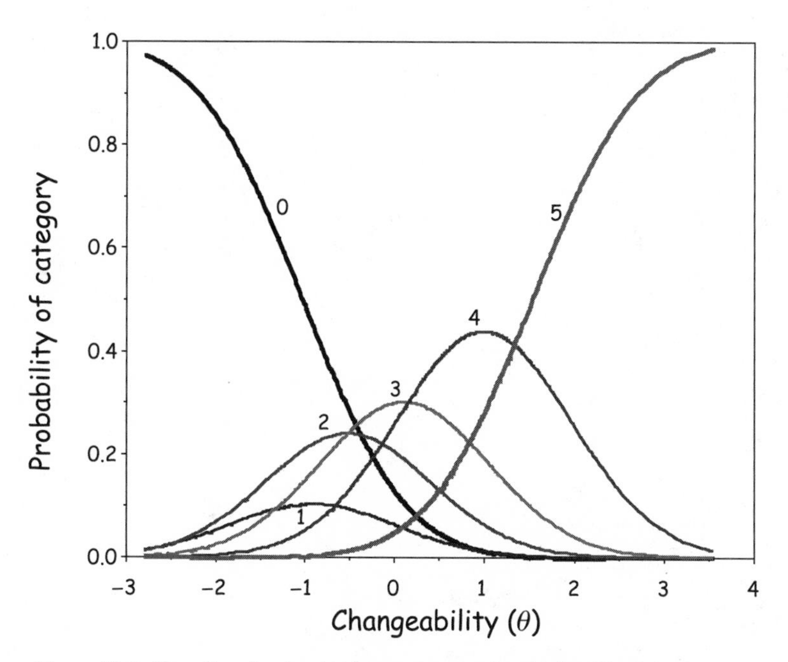

Figure 11.6. Tracelines for the six response categories obtained for Item 1.

discussion, we point out the extent to which methods that fail to account for the structure of the data yield different results, and the substantive benefits of using the background information to answer the "whys." We use the same test scoring model for these data as we used in Example 1 – the testlet generalization of Samejima's (1969) polytomous model (as described in Chapter 10), with covariates embedded in a fully Bayesian framework.

11.3.2. The results

One standard inference that is obtainable from an IRT (or TRT) model for polytomous data are the probability curves (called tracelines) describing how, for each item, the probability of giving response category r varies with latent trait θ, in this case changeability. For example, the probability of each possible response, $r = 0, 1, \ldots, 5$, to Item 1 can be represented graphically as the six tracelines shown in Figure 11.6. From this, we can see that the responses to Categories 0, 1, and 2 are all close together, suggesting that these categories, as a function of latent changeability, do not really differentiate much among responders, whereas response Categories 3, 4, and 5 are better differentiated.

We derived similar figures for all 21 items. The study of these figures yields insights into the functioning of the survey instrument; items that have wider

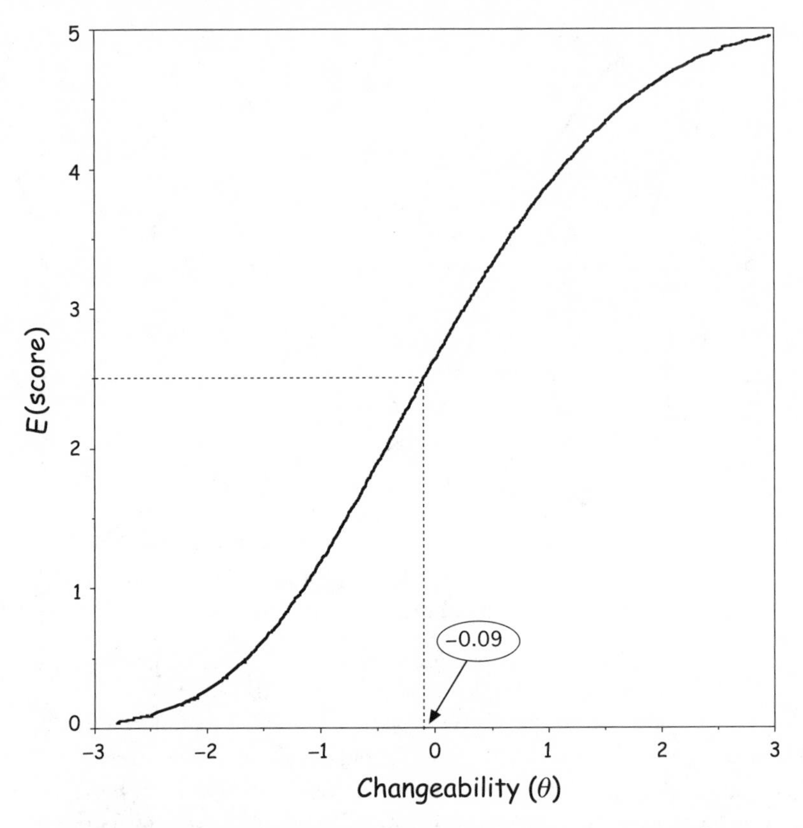

Figure 11.7. The expected score curve for Item 1 allows us to summarize an analog for difficulty for a polytomous item with the point on the x-axis at which the expected score reaches its 50% point.

separation indicate that respondents are consistently using the full range of response categories that are provided. In addition, they provide information on the impact of that item on the respondents.

Item tracelines also provide the raw material for an often useful summary. We can take the various component tracelines and combine them to yield an expected score curve $E(Y_j) = \Sigma_r r \times P(Y_j = r) = 0 \times P(0) + 1 \times P(1) + \cdots + 5 \times P(5)$. Such a curve for Item 1 is shown in Figure 11.7. We have drawn on Figure 11.7 a horizontal line at the 50% point of expected score (an expected score of 2.5) and indicated the value of the changeability score, $\theta = -0.09$, to which it corresponds. We see that a respondent whose changeability score is approximately 0 would have an expected response score to item j of about $r = 2.5$. Items that are less frequently cited as characterizing change would be offset to the right (e.g., a person would need to be more changeable to

Table 11.6. *The survey items and their 50% points on the Changeability scale*

Item no.	Item	50% point
3	An appreciation for the value of my own life	−0.58
11	Appreciating each day	−0.31
1	My priorities about what is important in life	−0.29
6	Knowing that I can count on people in times of trouble	−0.09
18	I learned a great deal about how wonderful people are	−0.05
12	Having compassion for others	−0.01
17	I discovered that I'm stronger than I thought I was	0.00
5	A better understanding of spiritual matters	0.01
8	Knowing I can handle difficulties	0.02
7	A sense of closeness with others	0.06
2	I'm more likely to try to change things that need changing	0.08
4	A feeling of self-reliance	0.18
10	Being able to accept the way things work out	0.24
16	I have a stronger religious faith	0.29
13	I'm able to do better things with my life	0.30
9	A willingness to express my emotion	0.36
15	Putting effort into my relationships	0.39
20	I accept needing others	0.56
21	I established a new path for my life	0.60
19	I developed new interests	0.75
14	New opportunities are available which wouldn't have been otherwise	1.03

endorse it) and those that are more frequently cited would be offset to the left.

An important technical aside illuminates expected score curves like that in Figure 11.7. Initially, one might be concerned that there is a loss of information when we move from the six tracelines in Figure 11.6 to their single summary curve in Figure 11.7. However, the expected score curve is uniquely determined by its component tracelines and uniquely determines those tracelines in turn (Chang & Mazzeo, 1994). Hence, no information is lost about an item using the expected score curve, making it an often used and useful summary.

Similar analyses were done for all of the items, and the 50% points thus obtained are shown in Table 11.6 along with an abbreviated version of the text of the items. Table 11.6 is ordered by the 50% points from the most highly endorsed to the least, and the items are spaced by leaving horizontal gaps in the table, based upon the apparent gaps in the item statistics (Wainer & Schacht, 1978). We see that the items that indicated the most change since being diagnosed

Testlet no.	50% point	Item no.	Testlet names
V	-0.6	3	Appreciation of life
	-0.5		
	-0.4		
V, V	-0.3	11, 1	
	-0.2		
I, I	-0.1	6, 18	
I, III, IV, III	0	12, 17, 5, 8	
I, II	0.1	7, 2	
III, III	0.2	4, 10	
IV, II	0.3	16, 13	
I, I,	0.4	9, 15	
	0.5		
I, II	0.6	20, 21	
II	0.7	19	New possibilities
	0.8		
	0.9		
II	1	14	

Figure 11.8. An annotated stem-and-leaf diagram makes clearer both the separation among survey items and the differences in the location of the testlets.

(3, 11, and 1) were all drawn from the "Appreciation of life" testlet. At the other end, we see that these patients with breast cancer acknowledge much less change in items that reflect "new activities" (19, 20, and 21). This structure is shown explicitly in the stem-and-leaf diagram in Figure 11.8. With these insights in hand, we now move on to describing the "whys."

11.3.3. Using covariates to understand why

We can use the covariates associated with each person as one explanatory information source to help us understand why some patients may have responded the way they did. Specifically, we might suspect that patients who were younger might have shown increased change due to their breast cancer diagnosis, as might those who have been diagnosed for a longer period of time. And what about the impact of the medical treatment that they received? Are those patients taking Tamoxifen more changeable? Or less?[3]

[3] This study was not designed to allow us to discover the direction of the causal arrow; for example, we don't know whether patients changed more because they took Tamoxifen, or whether the taking of Tamoxifen was engendered by a change in attitude.

Table 11.7. *Multiple regression of all covariates against changeability, as yielded from the survey*

Dependent variable is: CHANGEABILITY				$R = 0.38$
Source	Sum of squares	df	Mean square	F-ratio
Regression	116	8	14.5	12.9
Residual	668	597	1.1	
Variable	Coefficient	Standard error of coefficient	t-ratio	Probability
Constant	1.58	0.42	3.7	0.00
Age	−0.03	0.01	−5.8	≤0.0001
Tamoxifen	0.38	0.09	4.3	≤0.0001
Months since diagnosis	−0.04	0.02	−2.0	0.04
White	−0.19	0.10	−1.8	0.07
Working	0.16	0.11	1.5	0.13
Hispanic	0.16	0.14	1.2	0.24
Income	0.03	0.04	0.8	0.41
Married	−0.07	0.10	−0.7	0.51

To answer these questions, we next compare an analysis that uses regression-based methods to a Bayesian analysis where the covariates are directly included in the model, and inferences are derived directly within the MCMC framework described earlier. We highlight this distinction to emphasize the strength of including covariates directly as descriptors of the "whys." It also provides a caveat when doing post-MCMC regression analyses using point estimates as the dependent variable.

Using regression analysis

One traditional way to address "why" questions is to do a multiple regression with the parameter estimate of changeability (posterior mean) for each respondent as the dependent variable. A familiar kind of output from such an analysis is shown in Table 11.7. We discovered that these covariates have a modest, but statistically significant, relationship to the survey outcome (estimated latent changeability scores). A multiple R of 0.38 was found, indicating that about 15% of the total variance of the outcome is accounted for by the covariates. Following standard practice, we deleted those covariates whose relationship with the dependent variable was not statistically significant and recomputed the multiple regression with the reduced model. This reduced model accounts

Table 11.8. *Multiple regression of selected covariates against changeability, as yielded from the survey*

Dependent variable is: CHANGEABILITY				$R = 0.37$
Source	Sum of squares	df	Mean square	F-ratio
Regression	121	4	30.3	28.6
Residual	756	713	1.1	
Variable	Coefficient	Standard error coefficient	t-ratio	Probability
Constant	2.21	0.25	8.87	≤ 0.0001
Age	−0.04	0.00	−9.06	≤ 0.0001
Tamoxifen	0.33	0.08	4.20	≤ 0.0001
Months since diagnosis	−0.03	0.02	−1.88	0.06
White	−0.16	0.09	−1.90	0.06

for about 14% of the variance, justifying the removal of these insignificant variables, and is shown in Table 11.8.

Continuing with the traditional analysis, we would next interpret the covariates. Older women show less change compared with their younger counterpart; taking the drug Tamoxifen is associated with greater change; the longer the time since diagnosis, the less the expression of change, and White patients report less change than non-Whites. The other covariates are interpreted as not having any bearing on changeability because of the lack of statistical significance of their regression weights.[4] But such conclusions are often unwarranted and careful statisticians move much more carefully, deleting one variable at a time and examining how the pattern of regression coefficients changes, taking into account the intercorrelations among the independent variables and not deleting independent variables injudiciously. By being very careful, it is possible to arrive at the correct answer. Next, we report how a Bayesian approach gets us to the correct answer with far less genius required.

11.3.4. Using the posterior distributions of the covariate coefficients

The posterior distribution of a parameter is yielded automatically from a Bayesian analysis. It is the frequency distribution of the parameter draws, and its

[4] This method of interpreting the lack of statistical significance is often referred to as "the folly of, if it might be zero, it is."

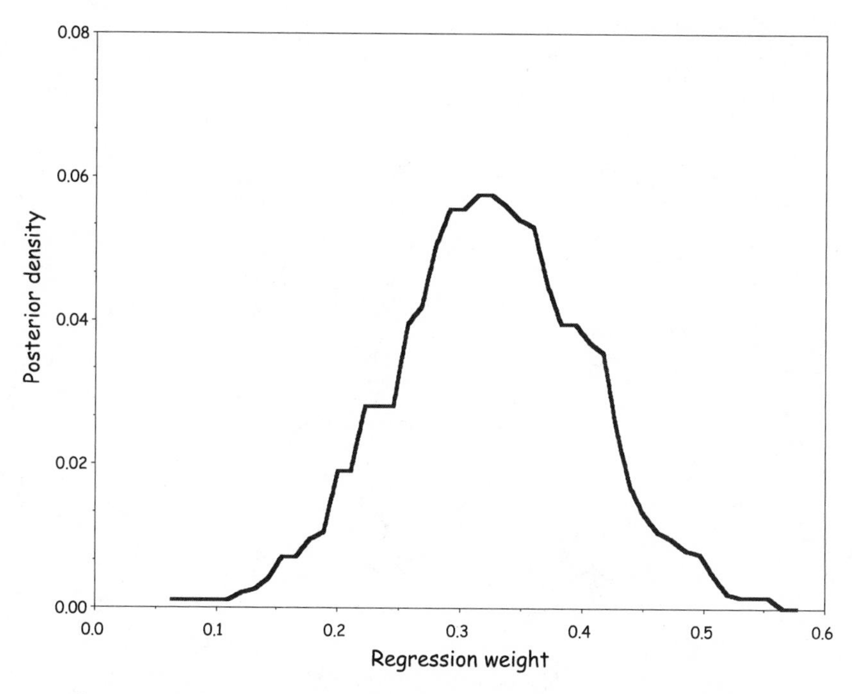

Figure 11.9. The posterior density of the regression weight on the covariate "Tamoxifen." Virtually all of the mass is above zero.

meaning is straightforward. We know that the point estimate of a parameter based on finite fallible data is not accurate. If we were to do the analysis again, different results would occur. This variation is due to the uncertainty involved in the sampling that is requisite in all data analysis. Although we know that the single answer that we chose to report is not the only possibility (which is why we always report some sort of error bounds around estimates), we usually pick the most likely value and report that (the value that maximizes the chance of observing the data we observed). The posterior distribution is merely the honest reporting of the whole distribution of true values with their associated likelihoods.

In Figure 11.9, we show the posterior distribution of the coefficient associated with the covariate "Took Tamoxifen." The frequentist analyses shown in Tables 11.6 and 11.7 strongly suggest that this is a significant predictor of changeability. Looking at the posterior distribution confirms the frequentist conclusion by showing us that virtually all the mass of the posterior lies above the zero value.

Examining the posterior distributions of the coefficients of the other three covariates that were judged "significant" shows a similar result.

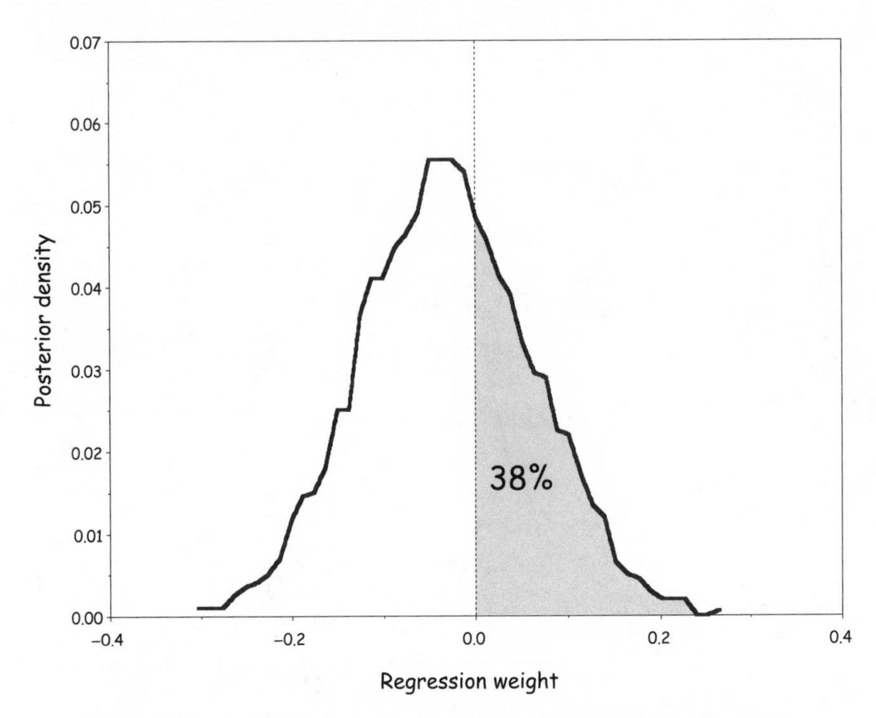

Figure 11.10. The posterior density of the regression weight on the covariate "Married"; 38% of the mass is above zero.

Next, let us look at the posterior distributions for two of the covariates (Hispanic and married) that were dismissed as not being significantly related to the outcome. Figure 11.10 shows the posterior density for the coefficient associated with the covariate "Married." We see immediately that the value of zero lies near the middle of the distribution, and hence we can conclude that marital status has little to do with a patient's responses to this survey. Again the Bayesian analysis supports the inferences we made from the frequentist regression.

Last, in Figure 11.11, we examine the posterior distribution for the covariate "Hispanic", and we see that 91% of the distribution lies to the right of zero. We would be amiss if we concluded that the event of being Hispanic was unrelated to survey responses. Here the conclusions from a Bayesian analysis are at odds with those drawn from the frequentist analysis. Of course, the frequentist analysis held hints that there might be something, but following standard practice would've eliminated it from the final model. A clever analyst might have been more judicious. This illustrates concretely Rubin's remark

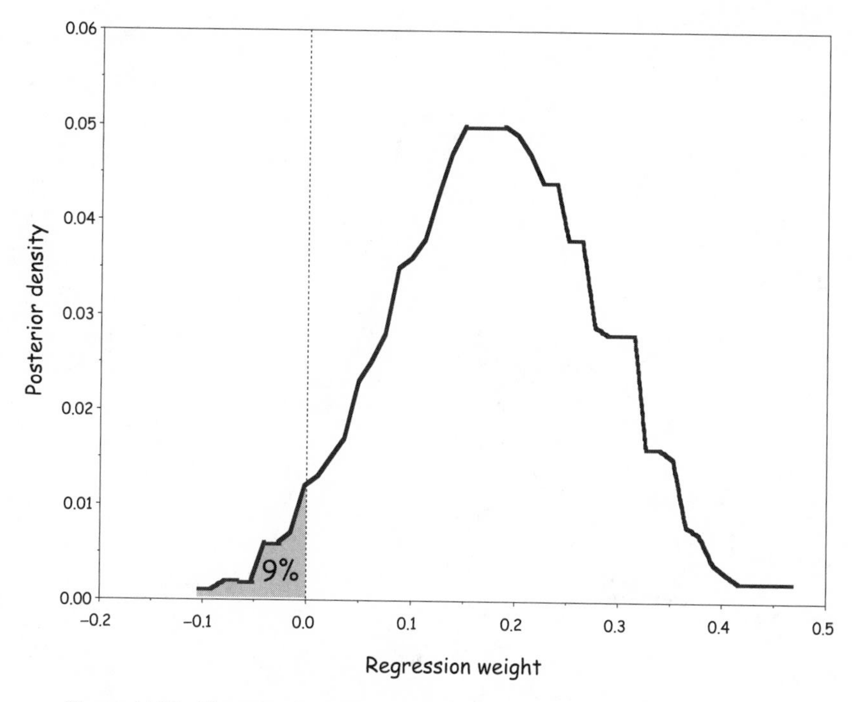

Figure 11.11. The posterior density of the regression weight on the covariate "Hispanic"; 91% of the mass is above zero.

quoted earlier about how Bayesian methods substitutes being careful for being brilliant in the quest for correct inferences.

The posterior distribution for "Working" closely resembles that for "Hispanic" and suggests including it in our inferences, whereas the posterior for "Income" resembles that for "Married" and can likely be excluded.

This illustrates how taking a Bayesian perspective allows us to make judgments directly from the results without recourse to arguments about standard errors and asymptotic normality.

11.3.5. What was the effect of local dependence?

Because the survey instrument was divided into five testlets, the model we fit allowed dependence within testlet. If these same data were to be fit with the analogous IRT model that assumes local independence, it is sensible to ask how different would the results have been. The answer, at least for point estimates of the parameters, is "not very different." This is reassuring, for IRT is often used

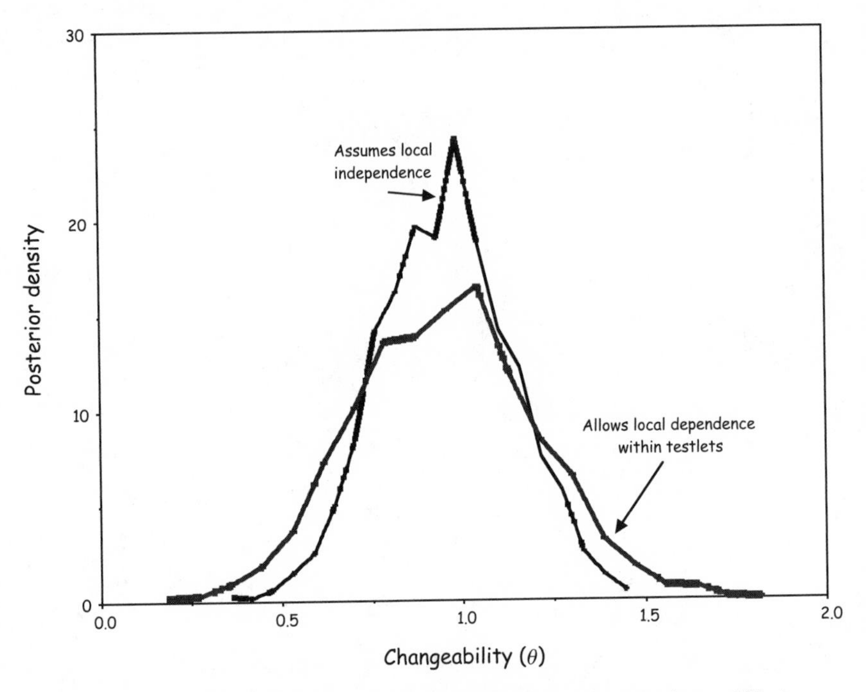

Figure 11.12. The posterior densities of the person parameter θ (changeability) for Person 1, shown for both models – the usual IRT model that assumes local independence, and the testlet model that allows within-testlet dependence.

when its assumptions are clearly being violated. The place where unmodeled local dependence affects results is in the estimates of parameter precision. If we assume independence when it is not true, we do not have as much information as we might've thought. As an extreme example, consider a test that consists of the same question repeated 20 times. If we assume that the responses to the 20 questions are independent, we would calculate a reliability measure on the basis of a 20-item test. But, in fact, we have a one-item test. A one-item test gives the same unbiased estimate of ability as a 20-item test, but the standard error of the estimate will be more than four times larger. This same result manifests itself with the current survey. In Figure 11.12, we show the posterior distributions of θ for Person 1 for two identical models, except one assumes local independence (IRT) and one does not (TRT). As is evident, they are both centered in the same place, but the TRT model is more platykurtic. The IRT model tells us that we have measured with greater precision than is, in fact, the case. For this survey, the variance of the posterior is underestimated by about 50% when averaged over all respondents.

11.3.6. Discussion

This example describes in some detail how the results from a survey instrument can be understood using the statistical machinery of testlet response theory. In a single coherent analysis, we have incorporated:

i. The desire to estimate each person's location on the latent dimension that underlies the 21 items on the survey.
ii. The testlet structure of the survey instrument.
iii. The covariates that were gathered to illuminate some of the reasons why the different respondents answered the way they did.

We found unsurprisingly that patients with breast cancer tell us that since their diagnosis they have become more focused on events and people close at hand and have considerably less interest in planning for the more distant future. We cannot say that the diagnosis is a cause of this response without a proper control group, but that would seem to be a plausible working hypothesis. We also cannot tell which direction the causal arrow is pointing after uncovering the relationship between taking Tamoxifen and reporting a change in attitudes. Such a causal link could be examined with longitudinal data. We do not know why there are ethnic differences in changeability, but the covariate analysis has alerted us to them for further consideration.

11.4. Conclusion

There are many possible sequels to the usual exercise of fitting a test scoring model to data and studying the results. In the past, the most common one was to report the scores to whoever was appointed to receive them. The item parameter estimates were usually sent to those technicians who were responsible for keeping track of the statistical characteristics of the test. At this point, the task of test data analysis typically ended. But sometimes analysts would go further and try to get past the questions akin to "what happened?" and move toward the more causal questions of "why?" Such extensions are rare enough that examples of wise investigations in this manner stand out in stark contrast. More than thirty years ago, Bock and Kolakowski (1973) looked at the distribution in a spatial visualization task and inferred a sex-linked recessive trait. Embretson (1984, 1995) used content information to try to understand what makes items difficult. The passage of No Child Left Behind tripled the amount of testing in the United States. This increase in testing had a concomitant increase in both quality and quantity of the inferences that the test scores were supposed to support. Answering "what" questions is no longer sufficient; we must be able to answer "why" questions as well. One part of the answer to "why" questions can be yielded through the judicious addition of covariates into the model. Covariates can help us understand what proportion of the variation in an item parameter we observe is accounted for by another observable variable (e.g., how much of the variation in the ability parameter is accounted for by differences in the number of years of training?). This chapter provides the methodology for including covariates as well as two illustrations of how it may be used.

Of course, one might have done similar analyses post hoc by regressing the model-based ability parameter against a set of independent variables, but this can lead to inaccuracies.

First, estimated relationships between TRT model parameters and the covariates of interest accurately reflect the uncertainty of the parameters that a post hoc regression with posterior means as the dependent variable would not.

A second advantage is that including covariate effects in the model reduces the unexplained sources of heterogeneity, allowing us to make more precise statements about the outcomes (e.g., we might originally say that the means of two subgroups are 1 standard deviation apart), but after adjusting for various observable characteristics of those groups (e.g., differences in age, coursework), we might find that the difference between them shrinks considerably. This might mean a different path toward remediation of the observed difference. An important technical point needs to be made explicit here – the multivariate process that we utilized for the item parameters allows for greater sharing of information across the TRT parameters, which then provides greater precision of estimation.

Finally, the contributions in this chapter represent the culmination of all of our methodological work in this arena over the past two decades. The product of this is a powerful tool that can provide a measurement scientist with sophisticated answers to difficult questions. The embedding of the test scoring model within a Bayesian framework provides an easy and accurate characterization of the stability of the model's parameters. The inclusion of covariate information lets us ask, and answer, "why?" in a straightforward way. In Chapters 13 and 14, we provide some additional illustrations of how these tools can be used.

Questions

1. Provide three examples of ways in which educational researchers would want to explain TRT model parameters via covariates. In each case, state the practical value of the findings.

2. Explain under what conditions covariate effects can be taken as causal versus observational.
3. Write down how the most general model, as described here, simplifies to the non-Bayesian versions of the 1-PL model, 2-PL model, 3-PL model, and Samejima ordinal response model.

References

Bock, R. D., & Kolakowski, R. (1973). Further evidence of sex-linked major-gene influence on human spatial visualizing ability. *American Journal of Human Genetics, 25*, 1–14.

Chang, H., & Mazzeo, J. (1994). The unique correspondence of the item response function and the item category response functions in polytomously scored item response models. *Psychometrika, 59*, 391–404.

Embretson, S. E. (1984). A general multicomponent latent trait model for response processes. *Psychometrika, 49*, 175–186.

Embretson, S. E. (1995). A measurement model for linking individual change to processes and knowledge: Application to mathematical learning. *Journal of Educational Measurement, 32*, 277–294.

Gelman, A., Carlin, J. B., Stern, H. S., & Rubin, D. B. (1995). *Bayesian data analysis.* London: Chapman & Hall.

Holland, P. W., & Wainer, H. (1993). *Differential item functioning*. Hillsdale, NJ: Erlbaum.

Samejima, F. (1969). *Estimation of latent ability using a response pattern of graded scores* (*Psychometrika Monographs* Whole No. 17). Richmond VA: Psychometric Society.

Spearman, C. (1904). "General intelligence" objectively determined and measured. *American Journal of Psychology, 15*, 201–293.

Tedeschi, R. G., & Calhoun, L. G. (1996). The Posttraumatic Growth Inventory: Measuring the positive legacy of trauma. *Journal of Traumatic Stress, 9*(3), 455–472.

Wainer, H., & Schacht. S. (1978). Gapping. *Psychometrika, 43*, 203–212.

12

Testlet nonresponse theory: dealing with missing data

The discussion of test scoring models in the first 11 chapters of this book has been largely theoretical, in the sense that we assumed we were fitting either a complete data set or that the data that were missing were just like the data that we observed. Anyone who has ever had anything to do with a real data collection process knows that this is almost always a fiction. Yet, it is broadly agreed that such an initial approach is sensible. In physics, we begin assuming frictionless movement, simple harmonic oscillations, and nonturbulent flow. Then, once the basic ideas have been mastered, the methods are broadened to accommodate a closer approximation to the broad ramifying reality from which these simplified abbreviations had been so neatly abstracted. So too with testlet theory.

Real data sets are only rarely complete. Sometimes examinees accidentally omit an item, sometimes time runs out before the test can be completed, sometimes the examinee does not know an answer and elects to omit it, and sometimes the examiner elects to administer only a subset of the complete item pool to any specific examinee. In this chapter, we provide a formal way of characterizing missing data, which can help us to know better how we should deal with incomplete data sets.

In our consideration of the situations in which some of the test data are not observed, we suggest various strategies for appropriate inference. We also provide an indication of the effects on the magnitude and direction of the errors that missing data can induce on the validity of inferences that we might draw. We begin by considering traditional rectangular format data sets in which the rows of the data matrix represent examinee's responses for all items, and the columns represent different examinees responses for a particular item.

Let us assume that each examinee responds to the items in the order of the columns. Two different missingness patterns are easily recognized and are common. One is the situation with missing values sparsely and randomly observed throughout the columns. The other has a piling up of missing values toward the end of the rows. These two patterns are likely an indication of different mechanisms behind the missingness. For example, if the assumed missingness happens in the middle of the exam, it could be that this item was too difficult for that particular examinee and she did not know how to answer it; another possibility could be that this examinee just forgot to answer it. When missingness happens at the end of the exam, we begin to suspect that this examinee may not have had enough time to reach the end of the test.

Different missingness patterns affect the analysis of educational test data in different ways. If the unobserved responses occur because of an examinee's carelessness (e.g., forgetting to write down answers), such missing values are unlikely to be directly related to the construct being tested, and hence can be thought of as ignorable. Hence, inference could then proceed by just modeling the observed responses. More typically though, sparsely unobserved responses in the middle of an exam suggest that the items were outside of the examinee's knowledge and so she chose to omit them. Such missing values are related to examinee's proficiency relative to the difficulty of the item, and hence should be taken into account when scoring tests. The cause of the missingness for the "not reached" situation is more ambiguous. There could be several explanations. If the items were not reached because the previous items were too difficult for the examinee and hence she could not reach the end of a test, the missingness pattern is related to the ability being tested. If an examinee is not able to reach the items at the end because of other reasons unrelated to the construct being measured (e.g., slow reading speed in a test of mathematics), those missing values are not related to the ability being tested. Our point is that even within a specific pattern of observed missingness, one is rarely able to know with certainty what are the causes, and the causes are what matter.

Missingness patterns are crucial since the methods of dealing with the unobserved responses depend strongly on the nature of the missingness. In this chapter, we describe the different types of missingness patterns commonly encountered and link these to the statistical likelihood functions that have been described in previous chapters, introducing several corresponding methods. In addition, we describe how SCORIGHT deals with missing data through a simulated data set.

12.1. Missing at random and missing completely at random

Two central concepts in the missingness literature are missing at random (MAR) and missing completely at random (MCAR). We describe these briefly here. For a complete description, see Little and Rubin (1987).

Let $\mathbf{Y}$ denote the complete data and $\mathbf{y}_i = (y_{i1}, \ldots, y_{iJ})$ represent the i-th examinee's responses for the J items. We also need a missingness indicator $\mathbf{M} = (m_{ij})$ denoting the missingness for examinee i on item j, that is, $m_{ij} = 1$, if y_{ij} is missing, otherwise $m_{ij} = 0$. The different missingness patterns are characterized by the conditional distribution of $\mathbf{M} = (m_{ij})$ given $\mathbf{Y}$, say $f(\mathbf{M} \mid \mathbf{Y}, \phi)$, where ϕ are unknown parameters that govern the missingness process. If a missingness pattern, $\mathbf{M}$, does not depend on the values of the data $\mathbf{Y}$, missing or those that were actually observed, that is, if

$$f(\mathbf{M} \mid \mathbf{Y}, \phi) = f(\mathbf{M} \mid \phi), \text{ for all } \mathbf{Y}, \phi, \tag{12.1}$$

this missingness is called MCAR. This assumption does not assert that the pattern itself is random. Rather, it emphasizes that the missing pattern does not depend on an examinee's responses (observed or unobserved) to the test's items.

Because we cannot know $f(\mathbf{M} \mid \mathbf{Y})$ for unobserved $\mathbf{Y}$, and it may be that the likelihood of the data being observed is related to the value that would have been observed (a person who doesn't respond is more likely to have gotten the wrong answer, i.e., it is more likely that $y_{ij} = 0$ than $y_{ij} = 1$), the assumption of MCAR is not often considered credible; it is certainly not easily testable. Furthermore, in many cases it is a too restrictive assumption.

A less restrictive assumption is MAR, which states that the distribution of $\mathbf{M}$ depends only on the responses that are observed, not on those unobserved. If $\mathbf{Y}_{\text{ob}}$ denotes the observed responses and $\mathbf{Y}_{\text{miss}}$ denotes the unobserved responses (i.e., those for which $m_{ij} = 1$), then $\mathbf{Y} = (\mathbf{Y}_{\text{ob}}, \mathbf{Y}_{\text{miss}})$ and MAR means

$$f(\mathbf{M} \mid \mathbf{Y}, \phi) = f(\mathbf{M} \mid \mathbf{Y}_{\text{ob}}, \phi), \text{ for all } \mathbf{Y}_{\text{miss}}, \phi. \tag{12.2}$$

If the distribution of $\mathbf{M}$ depends on the missing values in the response data matrix $\mathbf{Y}$, the missingness pattern is Not Missing At Random (NMAR). For example, if an examinee omits item j ($m_{ij} = 1$) when he doesn't know the answer ($y_{ij} = 0$), the missingness is not at random because it depends on the value that he would have gotten had he answered it. These are obviously not independent.

Suppose the probability that an examinee responds correctly does not depend on the responses of other examinees and given his ability, each item score is conditionally independent of the others, that is the joint distribution of (y_i, m_i)

is indepedent across examinees and locally conditionally independent across items. Then, following the notation of Little and Rubin (1987),

$$f(\mathbf{Y}, \mathbf{M} \mid \boldsymbol{\theta}, \boldsymbol{\beta}, \phi) = \prod_{i=1}^{I} \prod_{j=1}^{J} f(y_{ij}, m_{ij} \mid \theta_i, \beta_j, \phi)$$

$$= \prod_{i=1}^{I} \prod_{j=1}^{J} f(y_{ij} \mid \theta_i, \beta_j) f(m_{ij} \mid y_{ij}, \phi), \qquad (12.3)$$

which holds if the following two assumptions are true:

1. $f(\mathbf{Y} \mid \boldsymbol{\theta}, \boldsymbol{\beta}, \phi) = f(\mathbf{Y} \mid \boldsymbol{\theta}, \boldsymbol{\beta})$ and
2. $f(\mathbf{M} \mid \mathbf{Y}, \boldsymbol{\theta}, \boldsymbol{\beta}, \phi) = f(\mathbf{M} \mid \mathbf{Y}, \phi)$.

In likelihood function (12.3), $\boldsymbol{\theta} = \{\theta_i, i = 1, \ldots, I\}$ denotes the proficiency of examinees, $\boldsymbol{\beta} = \{\beta_j, j = 1 \ldots, J\}$ denotes the item parameters, $f(y_{ij} \mid \theta_i, \beta_j)$ denotes the density of y_{ij} indexed by unknown parameters θ_i and β_j, and $f(m_{ij} \mid y_{ij}, \phi)$ is the density of a Bernoulli distribution for the binary indicator m_{ij} with probability $P(m_{ij} = 1 \mid y_{ij}, \phi)$ that y_{ij} is missing.

If Assumptions (1) and (2) hold, then the missing process does not depend on $\boldsymbol{\theta}$ or $\boldsymbol{\beta}$, and hence is *ignorable*. Thus, while there are still parameters ϕ from the distribution of $\mathbf{M}$ that could be estimated, they are distinct from the parameters of principal interest, those that govern the IRT model, and hence we can ignore $f(\mathbf{M} \mid \mathbf{Y}, \phi)$ when drawing inferences about $\boldsymbol{\theta}$ and $\boldsymbol{\beta}$.

Depending on which IRT model is being used, β_j could be a single item parameter or a vector of several item parameters. For example, for the 1-PL model, β_j is just the item difficulty parameter; for the 3-PL model, $\beta_j = \{a_j, b_j, c_j\}$ represents the item parameters of discrimination, difficulty, and guessing level. If missingness is independent of $\mathbf{Y}$, that is $P(m_{ij} = 1 \mid y_{ij}, \phi) = f(\phi)$, which does not depend on y_{ij}, then the missing pattern is MCAR, otherwise it is NMAR.

In general, when the missing data are MCAR, we could regard the observed data as a random subsample of the complete data. For example, suppose we have an item pool with N items in total and we randomly pick J items from this pool and assign them to I examinees. We can think of the resulting $I \times J$ data matrix as a random subset of the big $I \times N$ data matrix $\mathbf{Y}$ and then carry out the analysis based on the $I \times J$ data matrix knowing that such an analysis will be unbiased; however, because of its smaller sample size, it will have a larger standard error for all the estimates when compared with the same analysis based on the observations of the complete $I \times N$ data matrix. But, if the missingness is NMAR, the analysis based on the observed responses is likely to be biased for inferences about item parameters and examinee proficiency; however, it is possible for the data to be NMAR, but the missingness still to be ignorable for certain inferences.

An especially important instance of missingness is when there is a large item pool and each examinee is randomly assigned a subset of items from that pool. This is called a fractional or spiral design. Thus, each examinee can be administered different items, and hence potentially have different numbers of missing items as well as a different pattern of missingness. Such missingness still follows the definition of MCAR. In computerized adaptive tests (CAT), items are not assigned to each examinee at random, since item assignment depends on the examinee's responses to the previous items. However, because they are not dependent on the responses to items that have not been assigned, such a missingness pattern still follows the definition of MAR because it depends on the observed **Y**'s. This is what allows CAT to yield unbiased estimates of examinee proficiency despite the often enormous quantity of unasked and hence unanswered items.

An important counterexample to this (an example of NMAR) are ill-advised CAT item-selection algorithms (e.g., Lunz, Berstrom, & Wright, 1992) that allow examinees to reject items that are presented and force the algorithm to select another. Since the items that are responded to are determined both by the observed examinee responses and by the unobserved responses to the initially presented but ultimately rejected items, this process is NMAR and hence does not yield unbiased inferences. Moreover, examinations that allow examinees to select from several essays which one they would like to write have similar issues. For a more detailed discussion, see Bradlow and Thomas (1998) and Wainer and Thissen (1994).

12.2. Ignorable nonresponse

We saw in the previous section that one of the critical assumptions that allows us to obtain unbiased inferences based on the observed data is that the missing data mechanism is ignorable. The good news is that when nonresponse is considered to be ignorable, we can use methods of estimation on the basis of having complete data. For example, the choice of the 1-PL model, the 2-PL model, or the 3-PL model will each yield different likelihood functions that all apply to the case of incomplete data. The likelihood function, if we make the standard assumption of conditional independence of responses across persons and items, is just a product across those items for which the responses are observed. In this manner, we simply "skip over" the incomplete data and its indicators **M**.

In particular, we can write the likelihood function of $\mathbf{Y}_{\text{ob}}$ as

$$L(\boldsymbol{\theta}, \boldsymbol{\beta} \mid \mathbf{Y}_{\text{ob}}) \propto \prod_{i=1}^{I} \prod_{j \in J_{i,\text{obs}}} f(y_{ij} \mid \theta_i, \beta_j),$$

and correspondingly the loglikelihood function is

$$l(\boldsymbol{\theta}, \boldsymbol{\beta} \mid \mathbf{Y}_{\text{ob}}) \propto \sum_{i=1}^{I} \sum_{j \in J_{i,\text{obs}}} \log(f(y_{ij} \mid \theta_i, \beta_j)),$$

where $J_{i,\text{obs}}$ is the set of observed items for examinee i. Then, when a set of prior distributions $p(\boldsymbol{\theta}, \boldsymbol{\beta})$ are assigned to the proficiency of examinees and item parameters, inferences about $\boldsymbol{\theta}$ and $\boldsymbol{\beta}$ (after observing the data $\mathbf{Y}_{\text{ob}}$) are based on the observed data posterior distributions $p(\boldsymbol{\theta}, \boldsymbol{\beta} \mid \mathbf{Y}_{\text{ob}})$, which is determined by Bayes' rule:

$$p(\boldsymbol{\theta}, \boldsymbol{\beta} \mid \mathbf{Y}_{\text{ob}}) = \frac{p(\boldsymbol{\theta}, \boldsymbol{\beta}) L(\mathbf{Y}_{\text{ob}} \mid \boldsymbol{\theta}, \boldsymbol{\beta})}{p(\mathbf{Y}_{\text{ob}})}, \tag{12.4}$$

where $p(\mathbf{Y}_{\text{ob}}) = \int p(\boldsymbol{\theta}, \boldsymbol{\beta}) L(\mathbf{Y}_{\text{ob}} \mid \boldsymbol{\theta}, \boldsymbol{\beta}) d\boldsymbol{\theta}\, d\boldsymbol{\beta}$ is the normalizing constant.

The most common method for parameter estimation, on the basis of a likelihood function, is the maximum likelihood method (modal estimation) whose standard error and corresponding hypothesis tests are usually based on asymptotic (large sample) theory.

For Bayes inference, point estimates of $\boldsymbol{\theta}$ and $\boldsymbol{\beta}$ could be obtained from estimates of the center of the posterior distributions, such as the posterior mean or the posterior median (see Chapter 15 for a detailed discussion). To measure the posterior uncertainty, the posterior standard deviation replaces the maximum likelihood's standard error; the posterior probability interval (such as the 2.5-th to 97.5-th percentile of the posterior distribution or the 95% probability interval containing the highest values of the posterior density, called the highest posterior density (HPD) interval) replaces the confidence interval. Thus, the Bayesian confidence interval and the frequentist one measure two different things; the Bayesian measuring the uncertainty in the *parameter* of interest, whereas the frequentist interval is the uncertainty in the *estimator*. The p-value calculated by frequentists for hypothesis testing is similarly replaced by the posterior probability of the relevant event (the Bayesian tail area or p-value). However, if one chooses a constant as the prior distribution for all parameters, the Bayesian modal (maximum a posterior) estimates correspond to the maximum likelihood estimates. Of course, such a prior distribution is not a proper probability distribution.[1]

[1] However, if $p(\boldsymbol{\theta}, \boldsymbol{\beta} \mid \mathbf{Y}_{\text{ob}})$ is well defined, having improper priors still leads to valid inferences.

The goal of maximum likelihood estimation for complete and incomplete data is to find the values of $\boldsymbol{\theta}$, $\boldsymbol{\beta}$ that maximize the corresponding likelihood functions. If we let $f(\mathbf{Y} \mid \boldsymbol{\theta}, \boldsymbol{\beta}) = f(\mathbf{Y}_{\text{ob}}, \mathbf{Y}_{\text{miss}} \mid \boldsymbol{\theta}, \boldsymbol{\beta})$ denote the probability or density of the joint distribution of $\mathbf{Y}_{\text{ob}}$ and $\mathbf{Y}_{\text{miss}}$, the marginal probability density of $\mathbf{Y}_{\text{ob}}$ can be obtained by integrating out the missing data $\mathbf{Y}_{\text{miss}}$,

$$f(\mathbf{Y}_{\text{ob}} \mid \boldsymbol{\theta}, \boldsymbol{\beta}) = \int f(\mathbf{Y}_{\text{ob}}, \mathbf{Y}_{\text{miss}} \mid \boldsymbol{\theta}, \boldsymbol{\beta})\, d\mathbf{Y}_{\text{miss}}.$$

Finding the maximum likelihood estimators on the basis of the observed data is therefore equivalent to maximizing $f(\mathbf{Y}_{\text{ob}} \mid \boldsymbol{\theta}, \boldsymbol{\beta})$. Ignorable Bayesian inference for the parameters based on observed data $\mathbf{Y}_{\text{ob}}$ is obtained by incorporating a prior distribution for the parameters and basing inference on the posterior distribution

$$p(\boldsymbol{\theta}, \boldsymbol{\beta} \mid \mathbf{Y}_{\text{ob}}) \propto p(\boldsymbol{\theta}, \boldsymbol{\beta}) \times L_{\text{ign}}(\boldsymbol{\theta}, \boldsymbol{\beta} \mid \mathbf{Y}_{\text{ob}}), \tag{12.5}$$

where L_{ign} is the likelihood function of the observed responses.

If we use the missing indicator matrix $\mathbf{M}$, the full model treats it as a random variable. The joint distribution of $\mathbf{M}$ and $\mathbf{Y}$ is the product of the densities of the distribution of $\mathbf{Y}$ and the conditional distribution of $\mathbf{M}$ given $\mathbf{Y}$, which is indexed by an unknown parameter ϕ related to the missing pattern. That is,

$$f(\mathbf{Y}, \mathbf{M} \mid, \boldsymbol{\theta}, \boldsymbol{\beta}, \phi) = f(\mathbf{Y} \mid \boldsymbol{\theta}, \boldsymbol{\beta}) f(\mathbf{M} \mid \mathbf{Y}, \phi).$$

The full likelihood of $\boldsymbol{\theta}$, $\boldsymbol{\beta}$, and ϕ should be any function of $\boldsymbol{\theta}$, $\boldsymbol{\beta}$, and ϕ as follows:

$$\begin{aligned} L_{\text{full}}(\boldsymbol{\theta}, \boldsymbol{\beta}, \phi) &\propto f(\mathbf{Y}_{\text{ob}}, \mathbf{M} \mid \boldsymbol{\theta}, \boldsymbol{\beta}, \phi) \\ &\propto \int f(\mathbf{Y}_{\text{ob}}, \mathbf{Y}_{\text{miss}} \mid \boldsymbol{\theta}, \boldsymbol{\beta}) f(\mathbf{M} \mid \mathbf{Y}_{\text{ob}}, \mathbf{Y}_{\text{miss}}, \phi)\, d\mathbf{Y}_{\text{miss}}. \end{aligned} \tag{12.6}$$

When the missingness is MAR, and does not depend on the missing values $\mathbf{Y}_{\text{miss}}$, we can write

$$f(\mathbf{M} \mid \mathbf{Y}_{\text{ob}}, \mathbf{Y}_{\text{miss}}, \phi) = f(\mathbf{M} \mid \mathbf{Y}_{\text{ob}}, \phi).$$

Therefore,

$$\begin{aligned} f(\mathbf{Y}_{\text{ob}}, \mathbf{M} \mid \boldsymbol{\theta}, \boldsymbol{\beta}, \phi) &= f(\mathbf{M} \mid \mathbf{Y}_{\text{ob}}, \phi) \times \int f(\mathbf{Y}_{\text{ob}}, \mathbf{Y}_{\text{miss}} \mid \boldsymbol{\theta}, \boldsymbol{\beta})\, d\mathbf{Y}_{\text{miss}} \\ &= f(\mathbf{M} \mid \mathbf{Y}_{\text{ob}}, \phi) f(\mathbf{Y}_{\text{ob}} \mid \boldsymbol{\theta}, \boldsymbol{\beta}). \end{aligned}$$

And furthermore, if $\boldsymbol{\theta}$, $\boldsymbol{\beta}$, and ϕ are distinct, the maximum likelihood estimators for the parameters based on the $L_{\text{full}}(\boldsymbol{\theta}, \boldsymbol{\beta}, \phi \mid \mathbf{Y}_{\text{ob}}, \mathbf{M})$ will be the same as the maximum likelihood estimators based on the $L_{\text{ign}}(\boldsymbol{\theta}, \boldsymbol{\beta} \mid \mathbf{Y}_{\text{ob}})$, since the resulting likelihoods are proportional. It is essential for the IRT model parameters and the missing pattern parameters to be distinct. Examples exist in which this is a reasonable assumption and others for which it is not. For example, if the reason that data are missing is due to an ability not being tested, then the missing data mechanism can be considered ignorable for this inference.

Following the definition of an ignorable missing pattern in Little and Rubin (1987), we have:

Definition 1. The missing pattern is ignorable for likelihood inference if
(a) *MAR:* the missing data are missing at random; and
(b) *Distinctness:* the parameters of the likelihood model, $(\boldsymbol{\theta}, \boldsymbol{\beta})$, and the parameters that govern the missing data pattern, ϕ, are distinct.

For Bayesian inference, under the full model for $\mathbf{Y}$ and $\mathbf{M}$, the posterior distribution is obtained by combining the full likelihood (12.6) and a prior distribution for $(\boldsymbol{\theta}, \boldsymbol{\beta})$ and ϕ:

$$p(\boldsymbol{\theta}, \boldsymbol{\beta}, \phi \mid \mathbf{Y}_{\text{ob}}, \mathbf{M}) \propto p(\boldsymbol{\theta}, \boldsymbol{\beta}, \phi) \times L_{\text{full}}(\boldsymbol{\theta}, \boldsymbol{\beta}, \phi \mid \mathbf{Y}_{\text{ob}}, \mathbf{M}). \tag{12.7}$$

Suppose that the data are MAR, and

$$p(\boldsymbol{\theta}, \boldsymbol{\beta}, \phi) = p(\boldsymbol{\theta}, \boldsymbol{\beta}) p(\phi), \tag{12.8}$$

that is, $(\boldsymbol{\theta}, \boldsymbol{\beta})$ and ϕ are a priori independent. The posterior distribution is as follows:

$$\begin{aligned} p(\boldsymbol{\theta}, \boldsymbol{\beta}, \phi \mid \mathbf{Y}_{\text{ob}}, \mathbf{M}) &\propto [p(\boldsymbol{\theta}, \boldsymbol{\beta}) L(\boldsymbol{\theta}, \boldsymbol{\beta} \mid \mathbf{Y}_{\text{ob}})]\,[p(\phi) L(\phi \mid \mathbf{Y}_{\text{ob}}, \mathbf{M})] \\ &\propto p(\boldsymbol{\theta}, \boldsymbol{\beta} \mid \mathbf{Y}_{\text{ob}})\, p(\phi \mid \mathbf{Y}_{\text{ob}}, \mathbf{M}), \end{aligned}$$

that is, $(\boldsymbol{\theta}, \boldsymbol{\beta})$ and ϕ are posteriori independent. Bayesian inference about model parameters can therefore be based on the posterior distribution $p(\boldsymbol{\theta}, \boldsymbol{\beta} \mid \mathbf{Y}_{\text{ob}})$, ignoring the missing data. The corresponding definition of ignorable response from the Bayesian point of view is as follows:

Definition 2. The missing pattern is ignorable for Bayesian inference if
(a) *MAR:* the missing data are missing at random; and
(b) The model parameters and the missing parameter are prior independent, that is, the prior distribution has the form as given in (12.8).

Definition 2 is stronger than Definition 1 since the independence of the prior distribution implies the distinctness. In SCORIGHT, we deal with missingness by assuming that the prior distributions of $(\boldsymbol{\theta}, \boldsymbol{\beta})$ and ϕ are independent (we also assume that the priors for $\boldsymbol{\theta}$ and $\boldsymbol{\beta}$ are independent). Therefore, the prior distribution of $p(\boldsymbol{\theta}, \boldsymbol{\beta}, \phi)$ could be written as

$$p(\boldsymbol{\theta}, \boldsymbol{\beta}, \phi) = p(\boldsymbol{\theta}) p(\boldsymbol{\beta}) p(\phi).$$

By assuming local independence and treating missing data as MAR, SCORIGHT estimates the proficiency of examinees and item parameters on the basis of the posterior distribution $p(\boldsymbol{\theta}, \boldsymbol{\beta} \mid \mathbf{Y}_{\mathrm{ob}})$, ignoring the missing data.

12.2.1. An illustrative example

To show the effects of missing data, we generated a hypothetical data set with missing values containing 1,000 simulated examinees and 100 simulated items using the 2-PL logistic model. $\alpha\%$ missing data were generated as follows:

- We randomly generated 1,000 individual i-level n_i's (number of missing items) from a binomial distribution with upper bound $N = 100$ and probability of not responding $p = \alpha$.
- For each respondent i and their corresponding value of n_i, $i = 1, \ldots, 1{,}000$, we randomly generated a sequence of n_i numbers from 1 to 100 without replacement (the item index of missing responses from respondent i).
- For the i-th examinee, $i = 1, \ldots, 1{,}000$, we used the sequence of n_i numbers as the indicators for the missing response among those 100 items.

By such a mechanism, the missing pattern is MCAR. We choose $\alpha = 0.2$ and $\alpha = 0.4$ and analyzed four data sets with SCORIGHT: the complete one, one with 20% missing, one with 40% missing, and the 20% missing data set treating the missing observations as wrong answers.

In Figure 12.1, the three plots in the first row are the estimated values of the item discrimination parameter a by SCORIGHT versus the true values, the estimated item difficulty parameter b by SCORIGHT versus the true values, and the examinees' estimated proficiency θ by SCORIGHT versus the true proficiency values for the complete data set. The three plots in the second row are the same as those in the first row but with 20% of the data missing at random. The plots in the third row are for the data with 40% missing at random. The plots in the last row are the same as those in the second row (20% missing) but treat the missing observations as incorrect responses. Since SCORIGHT uses a fully Bayesian approach, the estimated values in the plots are the posterior means obtained from an MCMC sampler (see Chapter 15 for details).

For the results of the first three data sets, the plots reveal that the estimated values are all close to their true values, showing little bias. The estimates of item difficulty parameter b and examinees' θ's have less bias in comparison with the item discrimination parameter a. These results match with those obtained from a maximum likelihood method, for example, the results from BILOG (Mislevy & Bock, 1983) or MULTILOG (Thissen, 1991), and are entirely consistent with our experience using SCORIGHT to fit IRT complete data sets. If we look

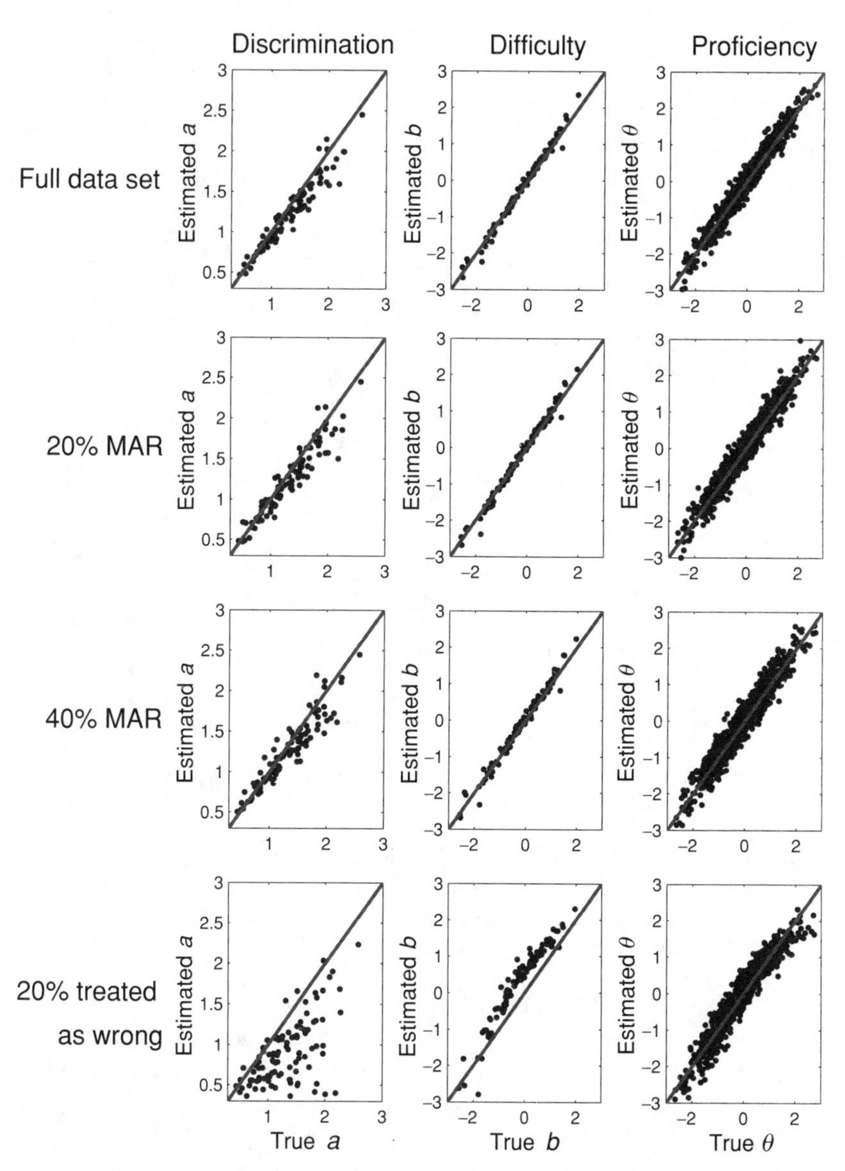

Figure 12.1. Analysis by SCORIGHT for the data sets: the complete one, the one with 20% missing at random, the one with 40% missing at random, and the one with 20% missing at random, but treating missing as wrong answers.

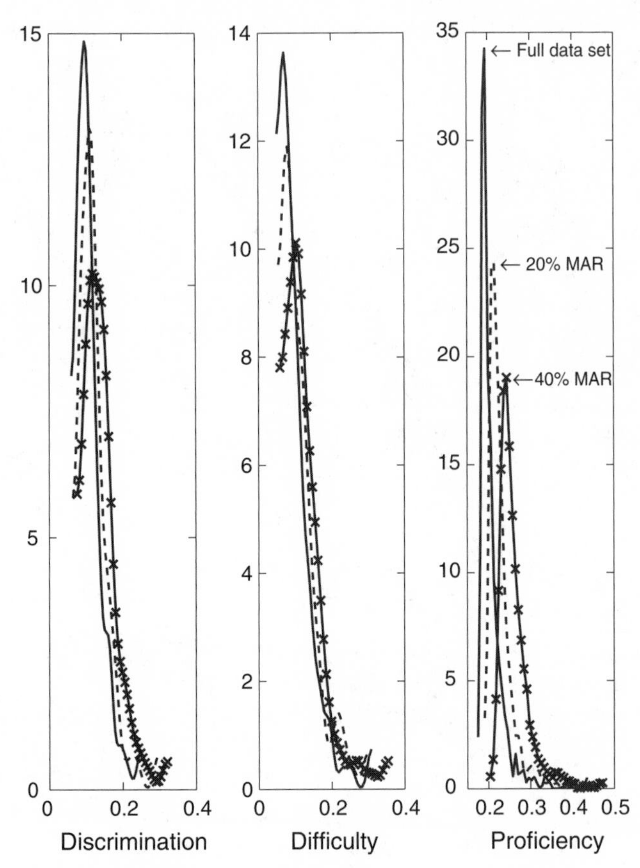

Figure 12.2. Kernel density estimation of the posterior standard deviation for estimated item parameter a, b, and examinees' proficiency θ for the three data sets: complete, missing 20%, and missing 40%.

at the plots in the last row, the estimates of item parameters and examinees' proficiency are off the diagonal lines, indicating that some bias was introduced by imputing the missing responses as wrong answers when, in fact, they were MAR. This is why we cannot endorse this last option unless a missing response, by design, is meant to be scored as incorrect.

If we look at the posterior standard deviations, we find, as expected, that they increase as the percentage of missing data increases. This is expected because here the percentage missing data is equal to the percentage missing information; albeit it is not true in general. Figure 12.2 shows the results of kernel density estimation for the posterior standard deviations of the 100 item parameter a's, b's, and the posterior standard deviation of the 1,000 examinees' proficiency θ's for the results of the first three data sets in Figure 12.1. The relationship between the mean of the posterior standard deviation and the percentage of missingness is clear. This simulation provides strong confirmatory evidence that the estimates from incomplete data when the missing elements, MCAR, are equivalent to the estimates based on a full data set with a smaller sample size; such missingness, when treated appropriately, does not cause bias but merely increases the posterior standard deviation proportional to the decrease in the sample size. The contrast with what was found in the last row of Figure 12.1 is stark; there, bias was created by the missingness mechanism and was not just an increase in the posterior standard deviation. Furthermore, the errors are compounded in that not only does treating the missing responses as incorrect introduce bias but also it incorrectly reduces the standard deviations to levels that would have been obtained had those observations been observed. Yet, it is common in most testing applications to treat missing responses as incorrect. If one does this but the data were, in fact, MAR, the missing data imputation scheme has introduced bias. Does this mean that we should treat missing test responses as MAR? To understand the consequences of this, and to see if this assumption is reasonable, we must delete observations in a decidedly nonrandom way and see what happens. We describe just such an experiment in section 12.3.

When SCORIGHT analyzes a test with testlet items (Chapters 6–10), it introduces a random effect $\gamma_{id(j)}$ to the linear predictor in the IRT model to tackle the violation of local independence. If the data are completely observed, and if we assume the 2-PL logistic model, the joint probability density function can be written as

$$\prod_{i=1}^{I}\prod_{j}^{J}\left[\frac{\exp(a_j(\theta_i - b_j - \gamma_{id(j)}))}{1+\exp(a_j(\theta_i - b_j - \gamma_{id(j)}))}\right]^{y_{ij}}\left[\frac{1}{1+\exp(a_j(\theta_i - b_j - \gamma_{id(j)}))}\right]^{1-y_{ij}}. \tag{12.9}$$

Since the $\gamma_{id(j)}$'s are parameters with distributions, they can be treated as missing data (see Little & Rubin, 1987). Thus, the complete data must be defined to include the unobserved $\gamma_{id(j)}$'s. If the original data matrix itself also has missing observations, then $\mathbf{Y}_{\text{ob}} = \mathbf{Y}_{\text{ob}}$ and $\mathbf{Y}'_{\text{miss}} = (\mathbf{Y}_{\text{miss}}, \{\gamma_{id(j)}\})$, where $\mathbf{Y}_{\text{miss}}$ represents the unobserved responses in the data matrix $\mathbf{Y}$ and $\mathbf{Y}'_{\text{miss}}$ represents the missing data under the model (12.9). SCORIGHT's treatment of missing data as ignorably missing under this situation is based on an additional assumption

$$f(\mathbf{M}\,|\,\mathbf{Y}, \gamma_{id(j)}, \phi) = f(\mathbf{M}\,|\,\mathbf{Y}_{\text{ob}}, \phi); \tag{12.10}$$

that is, ignorability conditional on the testlet effects. This assumption is what allows SCORIGHT to perform DIF analyses such as those to be described in Chapter 14. However, we discuss it briefly now as it is an application of missing data that "we cause"; and in a way that the reason that the missing data are MAR.

To test whether one item, say item j, behaves differently (perhaps of different difficulty) in two groups, which we call the focal group and the reference group, we duplicate this item and treat it as two items, item j and item j^*. We act as if the focal group has answered item j but omitted item j^*, and conversely the reference group answered item j^* and has omitted item j. SCORIGHT treats the missing observations as MAR and computes the posterior distributions of the parameters for both item j and item j^*. By examining the difference between the posterior distributions of b_j and b_j^*, one can tell whether this item has performed differently in these two groups.

When treating data as MAR, imputations for the ignorable missing responses can be accomplished using the EM algorithm (Dempster, Laird, & Rubin, 1977)[2] or using a fully Bayesian approach as described in other chapters. The advantage of the Bayesian approach is that it provides a natural imputation method on the basis of the posterior distributions that automatically yields an increase in the posterior standard deviation, rather than having to make adjustments that may be required through the use of EM. The extent of such an increase depends on the percentage of missing information.

12.3. Nonignorable nonresponse – omitted and not reached items

As discussed in Section 12.2, when the data are MAR and the distinctness of parameters ($\boldsymbol{\theta}$, $\boldsymbol{\beta}$) and ϕ exist, the likelihood based on the observed responses is proportional to the full likelihood with respect to the estimation of $\boldsymbol{\theta}$ and $\boldsymbol{\beta}$. Therefore, the maximum likelihood estimates and the Bayesian estimates of $\boldsymbol{\theta}$ and $\boldsymbol{\beta}$ based on the ignorable likelihood function are the same as the ones based on the full likelihood function L_{full}. In this situation, the estimation based on the marginal likelihood of the observed responses does not cause bias. The standard error of the estimates change depending on the percentage of missing information (this is not the same as the fraction of missing data (Little & Rubin, 1987), as shown in the examples of Section 12.2). However, when the missing pattern is NMAR, the full likelihood maximization requires a model of the missing indicators, **M**.

The selection of model for **M** depends on the analyst's knowledge. For test response data, if omitted items are treated as MAR, it means that the analyst believes that omitted items are caused by reasons unrelated to the examinees' ability or properties of the given missing items (the parameters of interest). However, in many cases, examinees may leave an item blank due to their inability to answer it correctly. Therefore, those missing values are not MAR and should be imputed using a model that takes the cause of the missingness into

[2] Technically, the EM algorithm imputes the sufficient statistics for the missing data, not the missing data.

account. The treatment of "not reached" items is an example of how the scoring of a test depends on the analysts' beliefs. If not reached items are caused by examinees' low proficiency, it might be reasonable to impute the missing data with some model or just treat them as wrong answers. If not reached items are caused by other effects, for example, slow reading speed for mathematics items, then the missing responses have nothing to do with the proficiency being tested, in which case, it is often reasonable to treat them as MAR.

If one utilizes a model for missingness that is incorrect, a bias would be induced. However, the better the imputation model, the greater the reduction of bias in the estimation that would be caused by assuming that the data are MAR. The following example illustrates how different imputing methods perform.

We continue with the general structure of the synthetic data set introduced in Section 12.2 but now the data are not MAR. The missingness is created probabilistically where those items that are less likely to be answered correctly are more likely to be missing. As before, the percentage of missing data for the whole data set is about 20%. The following analysis was done by SCORIGHT:

i. ignoring missing data, that is, treating the missingness process as ignorable;
ii. imputing the missing responses as wrong answers; and
iii. imputing the missing data as correct answers with probability proportional to an examinee's total score.

Figure 12.3 shows the estimates by i, ii, and iii versus the true values. It is clear that bias exists in all three cases, especially significant for item discrimination parameter a.

Table 12.1 shows the average of all the expected mean square errors based on the Bayesian inference provided by SCORIGHT:

$$(\text{Bayesian estimates} - \text{true value})^2 + \text{posterior variance}$$

for the data sets from the previous section that had data MAR and this section in which data are NMAR. Table 12.1 gives the results for seven different cases:

Case I: the complete data set,
Case II: the data set with 20% MAR,
Case III: the data set with 40% MAR,
Case IV: the same data set as Case II with imputing missing as wrong answers,
Case V: the data set with 20% NMAR,
Case VI: the data set in Case V imputing missing as wrong answers, and
Case VII: the data set in Case V imputing missing data according to the total score of each examinee.

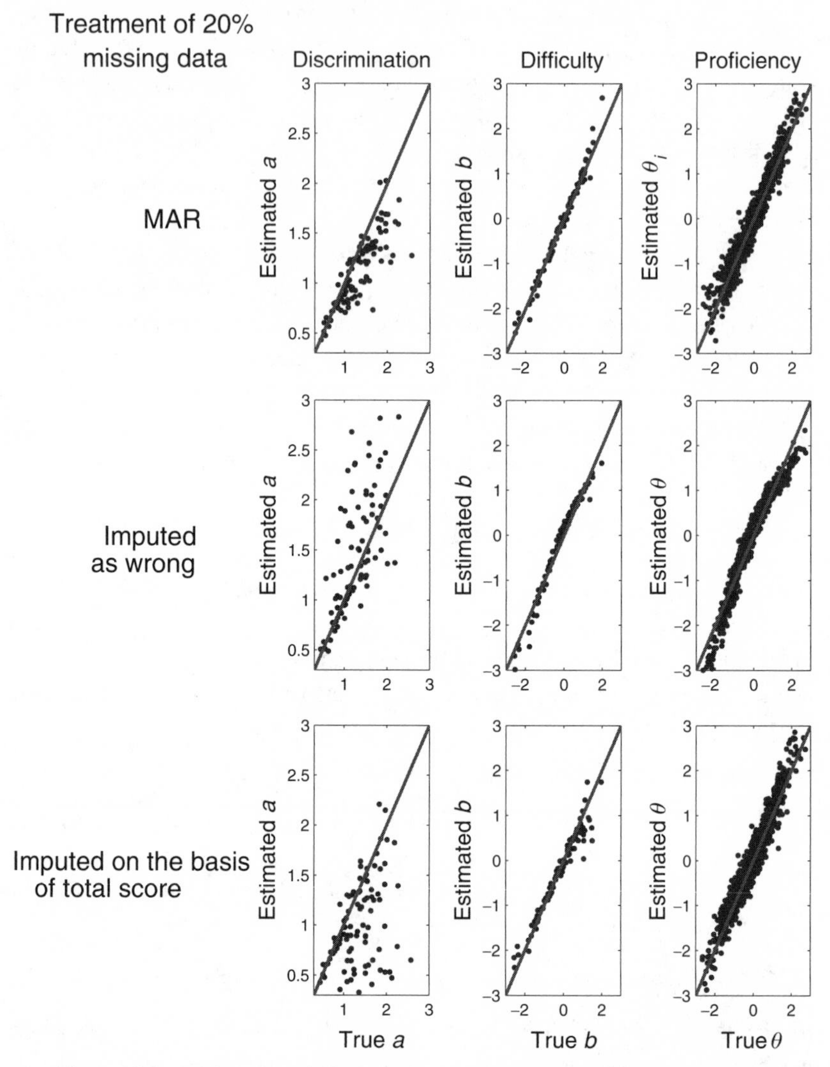

Figure 12.3. Analysis by SCORIGHT for the data sets simulated with NMAR data: (i) the incomplete observed data, (ii) completed data imputing missing data as incorrect, and (iii) completed data with imputed responses proportional to the total correct score for the examinee.

When missing data are ignorable, the expected mean square errors increase as the percentage of missing data increases. Their increases are due to the increase of the posterior variance just as we saw when data are MAR and the inferences were unbiased. If the imputing mechanism is wrong, as in Case IV, the bias

Table 12.1. *Expected mean square error (MSE)*

		a	b	θ
Case I	MSE	0.04	0.03	0.08
Complete	BIAS	0.03	0.02	0.04
	Variance	0.01	0.01	0.04
Case II	MSE	0.05	0.03	0.10
20% MAR	BIAS	0.03	0.02	0.05
	Variance	0.02	0.01	0.05
Case III	MSE	0.05	0.04	0.13
40% MAR	BIAS	0.03	0.02	0.07
	Variance	0.02	0.02	0.07
Case IV	MSE	0.33	0.35	0.15
20% MAR treated as wrong	BIAS	0.32	0.32	0.08
	Variance	0.01	0.03	0.07
Case V	MSE	0.13	0.09	0.12
20% NMAR	BIAS	0.11	0.07	0.06
	Variance	0.02	0.02	0.06
Case VI	MSE	1.32	0.05	0.09
20% NMAR treated as wrong	BIAS	1.25	0.04	0.06
	Variance	0.07	0.01	0.03
Case VII	MSE	0.34	0.20	0.13
20% NMAR imputed using θ	BIAS	0.33	0.18	0.07
	Variance	0.01	0.02	0.06

will be increased dramatically. For the nonignorable responses, Cases V, VI, and VII all give biased estimates. By imputing the missingness as incorrect responses, it improves the estimate of the item difficulty parameters b's and examinees' proficiency θ's, but it introduces a big bias for the estimates of item discrimination parameters a's, compared with the estimates based on the marginal likelihood. The imputation based on the total score does not provide any improvement in comparison with Case V. However, it does improve the estimates of the item parameter as compared with Case VI.

12.4. Conclusion

The principal lesson to be drawn from this chapter is that understanding the cause of the missingness process is crucial. If the process yields data that are NMAR, but you treat them as if they were, biased estimates would result. If the missingness process yields missingness that is at random, but you treat them as if there is a nonrandom pattern, bias would result as well. And, since the cause of missingness is rarely known, we are often operating in very dim light indeed.

That being said, data will have holes and we must do something. Happily there are instances where we are reasonably well-assured that the data are MAR. When this occurs, standard methods may be used. SCORIGHT, like most IRT programs, deals with missing data as if they were MAR.

But, when the data are NMAR, the only way that standard software can be used is to impute values into the holes of the data matrix before doing the final analysis. However, even though there is only one kind of MAR, there are many kinds of NMAR, consequently special attention needs to be given to the imputation process; different kinds of missingness will require different imputations. Under some circumstances, it is fair and proper for omitted items to be treated as "wrong" (e.g., if in the test instructions, examinees are told "omissions will be treated as incorrect"). Under other circumstances, this may not be so reasonable.

Missing data are ubiquitous, and hence always have to be dealt with one way or another, but whatever approach is taken will have consequences and possibly add bias to the outcome, so the decision about how to treat missing data should not be made capriciously. Moreover, it is usually wise to do sensitivity studies, akin to the simulations presented here, to measure the effect of the assumptions made about the character of the missingness mechanism. As ever, if one's conclusions vary greatly as a function of what is unknown, an alternative testing strategy that reduces missingness is probably the best option available.

References

Bradlow, E. T., & Thomas, N. (1998, Fall). Item response theory models applied to data allowing examinee choice. *Journal of Educational and Behavioral Statistics, 23*, (3), 236–243.

Dempster, A. P., Laird, N. M., & Rubin, D. B. (1977). Maximum likelihood from incomplete data via the EM algorithm. *Journal of the Royal Statistical Society, Series B, 39*, 1–38.

Little, R. J. A., & Rubin, D. B. (1987). *Statistical analysis with missing data*. New York: Wiley.

Lunz, M. E., Berstrom, B. A., & Wright, B. D. (1992). The effect of review on student ability and test efficiency for computerized adaptive tests. *Applied Psychological Measurement, 16*, 33–40.

Mislevy, R. J., & Bock, R. D. (1983). BILOG: Item analysis and test scoring with binary logistic models [Computer program]. Mooresville, IN: Scientific Software.

Thissen, D. (1991). MULTILOG user's guide (Version 6.1). Mooresville, IN: Scientific Software.

Wainer, H., & Thissen, D. (1994). On examinee choice in educational testing. *Review of Educational Research, 64*, 1–40.

PART III

Two applications and a tutorial

Thus far we have developed and discussed the machinery for models that can properly score tests that are made up of testlets. These models, especially the Bayesian ones discussed in Part II, are very general and hence there are many ways that they can be useful in the myriad situations that modern testing presents. Most standard test models nest within the general framework we have constructed. Traditional item response theory (IRT) emerges if all testlets contain but a single item; traditional true score theory devolves by treating a test as being made up of only one large testlet. In this part, we describe two special applications of these models as well as provide a tutorial on Bayesian estimation in general and MCMC methods in particular.

In Chapter 13, we examine the problem of measuring the efficacy of a test when there is a cut score. Traditional measures of reliability are incomplete, since they focus on the stability of the overall ordering of the examinees and not on the stability of the classification of examinees into classes of "passers" and "failers." We use this problem to illustrate how the posterior distribution of proficiency allows us to easily construct curves (posterior probability of passing or PPoP curves) that answer this question directly.

Chapter 14 deals with the differential performance of items and testlets in different examinee subgroups. Detecting and measuring differential performance has a rich history – both for items (differential item functioning, DIF) and for testlets (differential testlet functioning, DTF). In this chapter, we discuss how a Bayesian approach to this provides us with an answer that is more robust than the Mantel–Haenszel method for extreme circumstances.

Chapter 15 is a tutorial on Bayesian methods and estimation using Markov chain Monte Carlo (MCMC) procedures. The placement of this chapter posed a bit of a puzzle. According to one argument, it might have been the opening chapter for Part II; we opted instead to place it at the very end. We did this because it would have represented an interruption for readers already familiar with this widely used technology. It is written so that readers for whom this is new material can jump ahead to Chapter 15 for extra help without requiring that they know the material in the intervening chapters. We hope that this serves the purpose. It also provides extensive references to original source material for those who require more details.

13

Using posterior distributions to evaluate passing scores: the PPoP curve

13.1. Introduction

The goal of this chapter is two-fold. We want to provide an evocative example of how to use the posterior distributions of parameters and the samples from them that are the hallmark of a Bayesian approach to model fitting while solving a practical and often vexing problem in the analysis of test data. In this chapter, we show how to evaluate the efficacy of proposed passing scores for an exam. In Chapter 14, we use the posterior distributions in quite a different context, but with equal ease.

Passing scores are usually set through some sort of judgmental procedure (e.g., Angoff, 1971[1]). Once set, the issue immediately arises as to whether the test works well with that choice. For example, if a passing score is set well below the difficulty range of the existing pool of test items, accurate pass/fail decisions near the passing score cannot be made. It is at the passing score that the most critical decisions are made and where most of the scoring effort is placed (Bradlow & Wainer, 1998). What is needed is some way to assess how well the passing score and the test work together. The reliability of the test is a related idea but is more global. Test reliability gives us a measure of the quality of test as a whole, whereas a local measure near the set passing score is needed and is the purpose of this chapter; in fact, what we do is more general since we compute a local passing curve over the entire range of scores.

This type of reliability is often termed "classification consistency," and is generally defined as the extent to which decisions based on test scores from two independently administered, parallel forms of a test agree with one another.

[1] Angoff's method for standard setting is widely used, but its origin lies not with Angoff but with Ledyard Tucker, who provided the methodology in response to a problem posed to him by Angoff. Angoff never claimed credit for the development, only for writing it down; he pointed out clearly and often who was its true inventor. This is yet another instance of Stigler's Law of Eponymy (Stigler, 1980).

For many years, suggestions have been made concerning how to assess classification consistency (e.g., Cohen, 1960; Hambleton & Novick, 1973; Swaminathan, Hambleton, & Algina, 1974). Hambleton and Novick (1973) proposed a straightforward means: calculate the proportion of examinees consistently classified across two administrations and treat this as the reliability index for the cut-score decision. Swaminathan, Hambleton, and Algina (1974) elaborated this coefficient by correcting it for chance agreement, on the basis of Cohen's (1960) coefficient k.

These approaches are reasonable theoretically; but, in practice, data from two independent measurements for a particular examinee are rarely available. As a result, authors have proposed a variety of psychometric models to estimate classification consistency from a single test administration. The methods developed by Subkoviak (1976) and Huynh (1976) were based on the binomial and beta-binomial test models, respectively. Huynh (1990) considered classification consistency using the Rasch model. These methods required that test scores be composed only of dichotomously scored items. Livingston and Lewis (1995) extended such methods to allow for any scoring procedure, including weighted composite scores.

These methods were improvements in the assessment of classification consistency in that they produced robust estimates on the basis of a single test administration. However, such procedures followed the precedent set in the early work on classification consistency: reducing the reliability of a decision down to a single, easy-to-interpret coefficient. While such a solution is satisfying in some ways, it doesn't tell us anything about the conditional reliability of the decision being made; that is, how reliable is the pass/fail decision for various levels of examinee proficiency?

Some authors have approached this topic with considerations of conditional standard errors of measurement from various psychometric models (e.g., Kolen, Hanson, & Brennan, 1992). These approaches have been successful in telling us something about the conditional reliability of test scores, but provide relatively little information about the conditional consistency of classification decisions. Recent work using some strong true score models has addressed this concern (Lee, Hanson, & Brennan, 2002). On the basis of the assumptions of IRT and beta-binomial test scoring models, Lee, Hanson, and Brennan (2002) provide a means of constructing curves that represent the conditional probability of examinees being inconsistently classified by tests used to make multiple-category decisions. This approach has, as its spiritual forebear, the operating characteristic function of a statistical test, $P_\theta = P_\theta[D = H_1]$, where θ is a random variable and D is the decision function, and we are trying to decide between two competing hypotheses $H_0 : \theta = \omega_0$ or $H_1 : \theta = \omega_1$. This approach was initially described by Wald (1947) and more recently by Ghosh (1970). The method we describe here is a special case of the operating characteristic function, the posterior probability of passing (PPoP) curve. This approach is similar to the one outlined in Lee, Hanson, and Brennan (2002), with an important distinction. Unlike most strong true score models, the construction of PPoP curves does not require the assumption of a specific posterior density function for examinee proficiency.[2] In addition, it requires no additional work under the Bayesian paradigm, as the PPoP curve is readily available.

In the next section, we describe the idea behind the PPoP curve and how it is computed. In the third section, we apply this method to one part of the Clinical Skills Assessment (CSA), a test that is currently part of the U.S. licensing procedure for physicians who have been trained outside of the United States. Finally, we conclude with some remarks and suggestions for further uses of this methodology.

13.2. PPoP curves

A test and passing score that are well paired should yield the result that someone whose true score is well below the passing score will have a very low chance of passing the test, and similarly, someone whose true score is well above the passing score should have a very low chance of failing. Thus, it would seem sensible to measure the quality of the pass/fail decision by constructing a curve that relates the probability of passing to true score: the steeper the curve, the better the test score is working.[3] While it may seem natural to suggest that we should make the curve steeper everywhere, such an option is typically not feasible. How can we construct such a curve? One approach, not taken here (albeit sometimes sensible), is based on the maximum likelihood theory and the associated asymptotic variance.

Since we can never know what is a person's true score, we must obviously estimate it by something. A reasonable thing to use is the observed responses on the test. If the test is scored with a rigorous model (e.g., an

[2] This point requires a subtle distinction. The Bayesian inference is on $P(\theta >$ passing score$)$, whereas the frequentist reports the probability of the *estimate* of θ being greater than the passing score. This difference has important consequences for the assumptions that each has to make to get an answer. The frequentist approach typically requires that the estimators of θ are Gaussian; the Bayesian approach does not. Obviously, we are more interested in the probability that θ is greater than the passing score than we are in its estimate. However, if the posterior distribution of θ is indeed approximately normal as observed in many settings, as it was for most of the individuals in this research, then the results will be quite similar.

[3] This, of course, assumes that (a) the PPoP curve is monotone and (b) it is steeper in the right direction.

item response model), the person's true score is a parameter of that model. We can often obtain an asymptotic estimate of the standard error of that parameter estimate through the usual likelihood procedures (from the information matrix; Fisher, 1922, 1925), and then use these two estimates along with an assumption of normality to construct the probability of passing. Doing this for many score values and connecting them together with some sort of interpolated curve fit would yield a probability of passing curve. This is similar to the approach described in Lee, Hanson, and Brennan (2002). The validity of such a curve rests heavily on a variety of assumptions, including that the shape of the distribution of the estimated proficiency is Gaussian. When we are dealing with long tests (tests composed of many items), such an assumption is usually credible, but in other situations it may be suspect.

The Bayesian approach yields a much simpler calculation. The output of a Markov chain Monte Carlo (MCMC) estimation procedure are samples from the posterior distribution of the parameters of interest (see Chapter 15 for an extended discussion of MCMC methods). If that parameter is examinee proficiency, to calculate the probability of that examinee scoring above the passing score, all we do is draw a sample from the posterior and count how many times the sampled proficiency is above that passing score. If we extract 1,000 draws independently of one another (when using MCMC output, this is commonly accomplished by only recording every k-th draw, $k >> 1$) and 900 are above the passing score, we estimate the probability of that examinee surpassing that cut score as 0.90. To determine the PPoP curve, we merely sample examinees across the range of proficiency of interest, plot the passing probabilities, and connect the dots.[4]

13.3. Example: the integrated clinical encounter score of Clinical Skills Assessment

The Clinical Skills Assessment (CSA), instituted in July 1998, is one requirement for certification in the United States for foreign trained physicians. The assessment takes approximately 8 hours and involves 10 scored simulated encounters with actors (called "standardized patients" or SPs) who portray patients with a particular disorder. These encounters represent a cross-section of common reasons for patient visits to a physician. The candidates interview the

[4] More complex rules than simple linear interpolation may be called for when there are important parts of the score axis that are sparse.

SP and perform a focused physical examination. Following the encounter, the candidates summarize their findings in the form of a patient note, which is subsequently scored by a physician. In addition, the SP marks off, on a prepared scoring checklist, those questions that should have been asked and were. The number of items that are checked off becomes the candidates' "data gathering score" for that encounter; we treat this score as a polytomous response to the data-gathering question. Erroneously asked questions are simply ignored by the SP when scoring the candidate on the checklist. The data-gathering score and the patient note score are combined to yield the integrated clinical encounter score or ICE (for more details see Boulet et al., 1998). A second score, based on the effectiveness with which the candidate communicates with the SP (the communications score), completes the CSA. To pass the CSA, a candidate must achieve passing marks on both the ICE and communications subscores.

We fit the polytomous testlet model described in Chapter 10 to the ICE data for all examinees that took the CSA over the entire year beginning on June 29, 2001, until June 28, 2002. There were 6,066 examinees; 1,543 examinees spoke English as their first language, and the remaining 4,523 examinees did not. There were a total of 101 different clinical encounters, out of which each candidate was presented with 10.

The model we used for the CSA data is a parametric extension of the well-known Samejima IRT model (Samejima, 1969) for polytomous responses and is given by

$$P(y_{ij} = r) = \Phi(d_r - t_{ij}) - \Phi(d_{r-1} - t_{ij}), \tag{13.1}$$

but where in this instance we define the parameters of the probit link, t_{ij}, slightly differently; $t_{ij} = a_j(\theta_i - b_j - \gamma_{id(j)})$, where a_j and b_j are the usual item discrimination and difficulty for the jth item, θ_i is examinee i's proficiency, and d_r is the latent cut-off for score r. But, as we described in Chapter 10, the portion of this model that sets it apart from the traditional polytomous IRT models is the inclusion of the parameter $\gamma_{id(j)}$, the testlet interaction between examinee i and testlet $d(j)$, which contains item j. This term fits the excessive local dependence that characterizes testlets.

The model is constructed within a fully Bayesian framework and fitted using MCMC procedures operationalized in the computer program SCORIGHT. Of particular interest here is how we could use these results to study the efficacy of a passing score that was chosen through expert judgment. SCORIGHT outputs summaries of the posterior distribution of proficiency (such as the mean and standard deviation) and the draws themselves, which facilitates the computation that is described next.

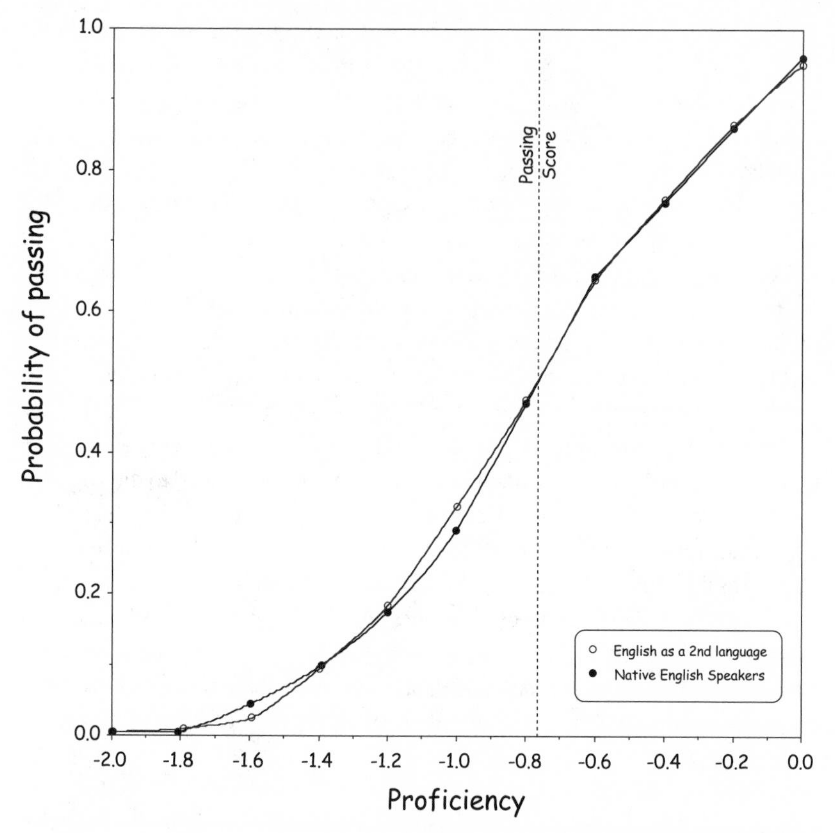

Figure 13.1. The probability of passing the ICE subtest as a function of proficiency for two language groups.

We used the posterior distributions from a selection of individuals whose proficiencies spanned the region around the passing score, and for each of them counted the proportion of draws from the posterior that were above the passing score. We did this separately for both native and nonnative English speakers. The results are shown in the PPoP curves in Figure 13.1. From this, we can see immediately that the ICE score functions essentially identically in the two language groups. In addition, the slope of the curve at the 50% probability of passing point is 2.2. This latter statistic becomes more meaningful when it is compared with other curves.

We redid the same analysis on the communication subscore and plotted the PPoP curves for both the communication and ICE subscores on the same axis. They are not designed to measure the same trait, but in both cases, the

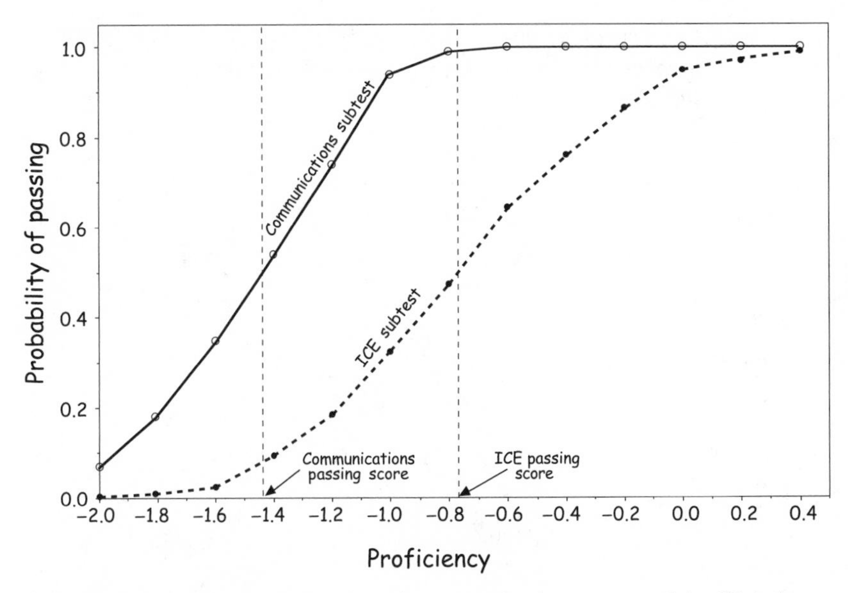

Figure 13.2. The probability of passing curves for the two parts of the Clinical Skills Assessment compared.

trait scores are scaled so that the population distribution of proficiency across individuals is normal with a mean of 0 and a variance of 1. Thus, since both scores are scaled in the same way, it is not unreasonable to look at their PPoP curves on the same plot. These are shown in Figure 13.2. From this display, we can tell immediately that the communication subtest is much easier (i.e., its passing score was set much below that of the ICE subtest) but its slope (2.6) at its passing score is steeper, indicating that it is better at discriminating among candidates whose performance is near its cut score.

The slope of the PPoP is roughly akin to reliability, but it measures the reliability of the decisions made on the basis of the test at that particular passing score. A natural question to ask is "how critical was the choice of passing score?" If the passing score were to be nudged in one direction or the other would the quality of the decisions change? This too is easy to answer from the samples from posterior distributions of proficiency. We simply construct PPoP curves for other passing scores. In Panel A of Figure 13.3 are the PPoP curves for the ICE subtest for five passing scores below the current passing score and five passing scores above the current one. These alternate choices are spaced 0.2 logits apart on the proficiency scale.

What we see immediately is that the choice of passing score could be made anywhere in a very broad range around the current value, with no profound

Panel A. An array of PPoP curves spanning more than two logits in location. Quality of passing decisions is robust with respect to choice of passing score.

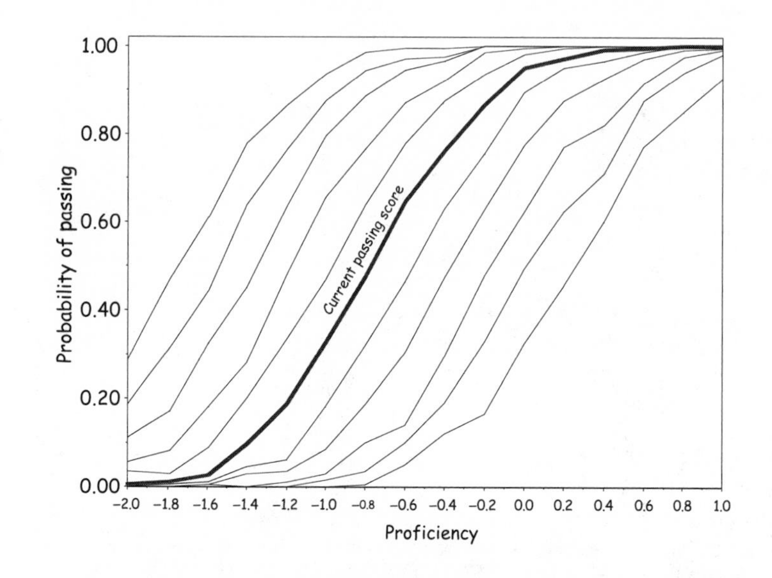

Panel B. Summary of PPoP slopes for the ICE score. The slopes of the entire array of PPoP curves are all about the same.

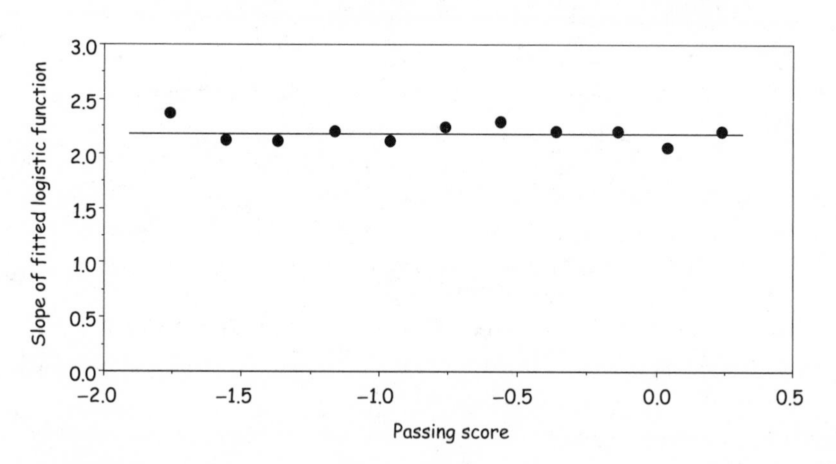

Figure 13.3.

change in the quality of the test's performance. This is shown explicitly when we plot the slopes of logistic functions fitted to these curves.

Of course, finding that the PPoP curves are approximately parallel should come as no surprise, because for the same person,

$$P(\theta > c) = P[(\theta + \text{constant})/\sigma_\theta > (c + \text{constant})/\sigma_\theta], \qquad (13.2)$$

and hence if the posterior distributions of all examinees' scores were the same, no matter where we looked (i.e., specifically $\sigma_\theta = \sigma$), we should see the structure we observed. What keeps the result in Figure 13.3 from being tautological is that we did not assume that the posteriors were all the same, but instead estimated them, and, in particular, the standard errors of the proficiencies. Thus, this result is a strong suggestion that more traditional procedures would probably work. Indeed, an interesting avenue for future research would be to learn how useful this parallelism in PPoP curves is in assessing the constancy of the standard error of measurement across the proficiency distribution.

One can also use the PPoP curves to calculate summary statistics that characterizes the accuracy of the test/cut-score combination. For example, we can calculate the false passing and false failing rates as well as their sum, the total error rate. To do this, we merely integrate the PPoP curve with respect to the ability distribution. More specifically, let us define $P(\theta)$ as the probability of passing, $F(\theta)$ as the ability distribution, and c as the passing score. Then, the probability of a false passing is $\int_{-\infty}^{c} P(\theta)\mathrm{d}F(\theta)$, and the probability of a false failing is $\int_{c}^{\infty} [1 - P(\theta)]\mathrm{d}F(\theta)$. The total error is the sum of these two.

In any finite sample, the integrals need only be replaced by summations over the population of examinees. When we do this for the communications subtest shown in Figure 13.2, we find that the probability of an incorrect pass is 0.028 and of an incorrect failure is 0.014. The total error rate is thus 0.042. But these two error rates need not be weighted equally; unequal weights are called for if the loss functions associated with each kind of error were not the same.

While it is clear that PPoP curves are a powerful and useful tool in understanding the quality of the passing decisions made with a particular test instrument, we have not demonstrated (or claimed) that a Bayesian approach is the only way to construct them. We could take the frequentist approach described in the Section 13.2, and use the maximum likelihood estimate of proficiency and its associated standard error in combination with a normal approximation and the hope that asymptotic results hold and produce much the same thing. Will this work?

A general answer to this question is grist for further research, but for these data, the difference between the two approaches was small, and so in this one instance the Gaussian approximation would be satisfactory. But, even when both methods work, our preference is the Bayesian approach; it is straightforward and easily allows many other kinds of generalizations, for once you have the posterior distributions, any sort of stochastic statement could be made just by counting.

13.4. Discussion and conclusion

In this chapter, we have provided an approach for assessing the efficacy of a cut score in a pass/fail examination at any point along the proficiency continuum. This method is, in its essence, graphical. One could easily derive some sort of summary statistic, such as the slope of the PPoP curve at its middle value, but we believe that such summaries may omit valuable information. One goal of evaluation is uncovering the unexpected, and often the best way to accomplish this is graphical. Our approach represents the nexus of three contemporary trends in thoughtful data analysis: the graphical display (Wainer, 2000, 2005) of functions (Ramsay & Silverman, 1997) derived within a Bayesian framework (Gelman et al., 1995). Furthermore, given the increasing popularity of Bayesian methods in test scoring, PPoP curves emerge almost for free.

Future research in this area, given the computational demands of MCMC methods, is certainly needed as to when the methods described here would provide the biggest payoff. Of course at some level, the answer to this question depends upon both the loss function for the two kinds of errors and the number of examinees whose proficiency is near the cut score. Either way, we believe this is a useful method that should be in the toolkit of the applied educational researcher.

We cannot end without adding a small word on behalf of the Bayesian methods that, with the development of Monte Carlo estimation methods such as the Gibbs sampler, have recently become practical. Using a Bayesian approach requires care and computer time, but not genius.

Questions

1. What's a PPoP curve?
2. How is a PPoP curve different from a frequentist alternative?
3. What role does the posterior distributions of the model parameters play in the construction of PPoP curves?

References

Angoff, W. H. (1971). Scales, norms and equivalent scores. In R. L. Thorndike (Ed.), *Educational measurement* (pp. 508–600). Washington, DC: American Council on Education.

Boulet, J. R., Ben David, M. F., Ziv, A., Burdick, W. P., Curtis, M., Peitzman, S., & Gary, N. E. (1998). Using standardized patients to assess the interpersonal skills of physicians. *Academic Medicine, 73*, S94–S96.

Bradlow, E. T., & Wainer, H. (1998). Some statistical and logical considerations when rescoring tests. *Statistica Sinica, 8*, 713–728.

Cohen, J. (1960). A coefficient of agreement for nominal scales. *Educational and Psychological Measurement, 20*, 37–46.

Fisher, R. A. (1922). On the mathematical foundations of theoretical statistics. *Philosophical Transactions of the Royal Society of London A, 222*, 309–368.

Fisher, R. A. (1925). Theory of statistical estimation. *Proceedings of the Cambridge Philosophic Society, 22*, 700–725.

Gelman, A., Carlin, J. B., Stern, H. S., & Rubin, D. B. (1995). *Bayesian data analysis.* London: Chapman & Hall.

Ghosh, B. K. (1970). *Sequential tests of statistical hypotheses*. Reading, MA: Addison-Wesley.

Hambleton, R. K., & Novick, M. R. (1973). Toward an integration of theory and method for criterion-referenced tests. *Journal of Educational Measurement, 10*, 159–170.

Huynh, H. (1976). On the reliability of decisions in domain-referenced testing. *Journal of Educational Measurement, 13*, 253–264.

Huynh, H. (1990). Computation and statistical inference for decision consistency indexes based on the Rasch model. *Journal of Educational Statistics, 15*(4), 353–368.

Kolen, M. J., Hanson, B. A., & Brennan, R. L. (1992). Conditional standard errors of measurement for scale scores. *Journal of Educational Measurement, 29*, 285–307.

Livingston, S. A., & Lewis, C. (1995). Estimating the consistency and accuracy of classifications based on test scores. *Journal of Educational Measurement, 32*, 179–197.

Lee, W-C., Hanson, B. A., & Brennan, R. L. (2002). Estimating consistency and indices for multiple classifications. *Applied Psychological Measurement, 26*, 412–432.

Ramsay, J. O., & Silverman, B. W. (1997). *Functional data analysis.* New York: Springer-Verlag.

Samejima, F. (1969). *Estimation of latent ability using a response pattern of graded scores*. (Psychometrika Monographs, Whole No. 17) Richmond, VA: Psychometric Society.

Stigler, S. M. (1980). Stigler's Law of Eponymy. *Transactions of the New York Academy of Sciences, 2nd Series, 39*, 147–157.

Subkoviak, M. (1976). Estimating reliability from a single administration of a criterion-referenced test. *Journal of Educational Measurement, 13*, 265–275.

Swaminathan, H., Hambleton, R. K., & Algina, J. (1974). Reliability of criterion-referenced tests: A decision-theoretic formulation. *Journal of Educational Measurement, 11*, 263–268.

Wainer, H. (2000). *Visual revelations: Graphical tales of fate and deception from Napoleon Bonaparte to Ross Perot* (2nd ed.). Hillsdale, NJ: Erlbaum.

Wainer, H. (2005). *Graphic discovery: A trout in the milk and other visual adventures.* Princeton, NJ: Princeton University Press.

Wald, A. (1947). *Sequential analysis.* Wiley: New York.

14

A Bayesian method for studying DIF: a cautionary tale filled with surprises and delights

14.1. Introduction

It is a truth universally acknowledged, that an educational researcher in possession of a good data set, must be in want of a Bayesian method.

(Paraphrased, with our apologies, from Jane Austen, 1813)

Traditional, frequentist-based, statistical methods have provided researchers with a lens through which they could fruitfully examine the world. These well-known methods for both continuous (e.g., regression) and discrete outcome data (e.g., contingency table procedures) have proved their worth innumerable times. Yet, all methods have limits of applicability and these are no exceptions to that rule. The growth of analogous robust procedures have helped to provide the unwary analyst with protection against unexpected events – usually longer-tailed-than-Gaussian error distributions – but despite the proven robustness of these methods, careful thought must accompany any analysis. If one blindly utilizes statistical methods even those as well established as the Mantel–Haenszel (MH) test considered here, one does so with some risk. The computationally intensive procedures that we have used in fitting the Bayesian testlet models are continually becoming "cheaper and cheaper" and have made a Bayesian approach to many problems a practical lens through which phenomena can be viewed. Such methods provide further protection against the unknown. Remember Donald Rubin's comment, quoted in Chapter 7, that

> If you are very smart you can use frequentist methods to solve hard problems, but you can be dumb as a stump and still get the right answer with Bayesian methods.
>
> *(Rubin, personal communication, October 26, 2004)*

In this chapter, we provide a vivid illustration of this in our study of differential item functioning (DIF).

As we described in Chapter 6, *DIF* is a term used to describe a test item that behaves differently in different subgroups of the examinee population. It is distinguished from *impact* in that the latter term does not adjust for differences in the proficiency of the two examinee groups being compared. For instance, a particular test item may behave differently for U.S. and foreign-born examinees (hence the item has impact), but not for U.S. and foreign-born examinees of the same estimated ability level (hence no DIF). An encyclopedic discussion of methods for uncovering DIF can be found in Holland and Wainer (1993). Although there have been many methods developed for uncovering and studying DIF, by far the most widely used is an approach developed by Holland and Thayer (1988) in which they adapt the MH statistic from its original use in cancer research (Mantel & Haenszel, 1959). The MH test stratifies the examinee population by some measure of proficiency, usually total score on the test (or, more generally, estimated ability[1]), and within each stratum calculates an odds ratio on the basis of a two-way table of counts that has, as rows, examinees in the reference population and examinees in the focal population, and as columns, those who got the item correct and those who got it wrong. The MH test then summarizes all of these two-way tables (one for each stratum) by summing them appropriately to form the MH test statistic. The differential odds ratios of correct/incorrect across different ability levels would be an indication, by definition, of DIF.

The MH was originally proposed for use in DIF studies for three principal reasons:

i. It was a tried-and-true method that had been in use for more than thirty years, and in that time had developed a reputation as a reliable and accurate measure.
ii. It focused all of its statistical power on estimating a single parameter, and hence was especially appropriate when the goal of the researcher was to be able to accept an item after finding insufficient evidence to declare that the item was performing differentially in the two populations.
iii. It could be calculated quickly and easily, and hence could be done within the tight practical constraints under which commercial testing programs usually labor.[2]

[1] Total score is a sufficient statistic for ability in the 1-PL and the Rasch model. When more general models are used, people tend to stratify on estimated ability. A somewhat open issue is whether the estimated ability should or should not include the focal item when testing for DIF using the MH statistic.

[2] Paul Holland characterized this property at "10¢ an item". He felt that so long as he could keep the cost of doing DIF under this rate, test program directors would have one less excuse for not checking for DIF. Recently he updated his cost estimate to "1¢ an item."

In the almost twenty years since the MH test has been proposed for screening test items for DIF, there have been an enormous number of research studies, usually simulations, that have tried to uncover the bounds within which it could safely be used (e.g., Donoghue, Holland, & Thayer, 1993; Clauser, Mazor, & Hambleton, 1991, 1993, 1994; Mazor, Clauser, & Hambleton, 1992, 1994). A typical study done by Rogers and Swaminathan (1993) tested for Type I and Type II errors under a range of conditions, and generally found that the MH test statistic's performance was impeccable. Among their conditions were items with a range of difficulty (b between -1.5 and $+1.5$) as well as a range of discriminations (a between 0.6 and 1.6). In many testing situations, this covers a large range of the potential items of interest; however, they did not cover one common testing situation.

We know of no studies that have tested the efficacy of the MH test at truly extreme values of difficulty and discrimination (e.g., $b < -5$ and $a < 0.4$). Yet, such extreme values occur more often than seldom in licensing examinations (which tend to have a strong criterion referenced flavor) in which the examinee population is exceptionally able. This is an apt description of medical licensing, and hence this is the scenario we consider here when suggesting, and providing the method for, a more careful approach for detecting DIF. In particular, via Bayesian DIF methods, we provide a second lens through which to study item or testlet performance.

In Section 14.2 we provide a Bayesian approach to studying DIF that is based on testlet response theory (TRT) and is directly analogous to the well-known likelihood ratio methods for DIF detection. In Section 14.3, we describe a data set drawn from the third part (Step 3) of the United States Medical Licensing Examination (USMLE™) that provides an empirical illustration of our method. Section 14.4 contains the results obtained from both the MH and the Bayesian DIF procedures, and in Section 14.5, we discuss these results as well as provide the results of two simulation studies we ran to confirm our findings. Section 14.6 contains some concluding advice for those interested in DIF analyses and areas for future research.

14.2. A Bayesian procedure for measuring DIF

The method we propose relies on the estimation methodology used for the testlet response model (and TRT) introduced and described in Chapters 7 to 12 and implemented in the computer program SCORIGHT. The approach we espouse here uses the more general testlet model rather than the standard item response theory (IRT) models for four primary reasons. First, most examinations today contain items that are nested within testlets (e.g., a single reading passage "stem" with multiple questions based on it) and ignoring such dependence could lead to biased parameter estimates. Second, the model can be specialized to testlets with but a single item, which is isomorphic to the usual IRT model that assumes conditional independence between items. Hence, there is no "cost" in using TRT as it nests IRT as a special subcase. Third, the data set we analyze here has a robust testlet structure, and hence the more general model is perfectly appropriate. Finally, the availability of Bayesian TRT software (SCORIGHT) and increased computing power has made its implementation feasible and no more difficult than Bayesian IRT.

14.2.1. The model

We begin by recapitulating the basic characteristics of the TRT model utilized here.

1. The TRT model can handle a mixture of both polytomous and binary response items nested and not nested within testlets. As the data set here considers binary items only, we describe only the binary portion of TRT, although the generalization to testlets should be obvious.
2. The model fit to binary items is the 2-PL model given by $P(y_i = 1 \mid t_{ij}) = \text{logit}^{-1}(t_{ij})$, where t_{ij} is the latent linear predictor of score, $y_{ij} = 0$ or 1, the response score from person i on item j, and $\text{logit}^{-1}(t_{ij}) = \exp(t_{ij})/(1 + \exp(t_{ij}))$. As the 2-PL model is a widely used model, it serves well as the standard model. Extensions to the 3-PL TRT model were given in Chapter 9.
3. The assumed model for t_{ij} was extended from the standard 2-PL IRT model in which $t_{ij} = a_j(\theta_i - b_j)$ to one that incorporates testlet dependence, TRT, – $t_{ij} = a_j(\theta_i - b_j - \gamma_{id(j)})$, where as before θ_i, a_j, and b_j are the person ability, item discrimination, and item difficulty, respectively. The additional term in the model, $\gamma_{id(j)}$, incorporates the extra dependence when two items j and j' are in the same testlet, that is, $d(j) = d(j')$.
4. The entire testlet model is embedded within a Bayesian framework that allows sharing of information across persons, items, and testlets. In particular, the structure that was employed here and implemented via SCORIGHT was
   i. $\theta_i \sim N(0, 1)$ to identify the model,
   ii. $[\log(a_j), b_j] \sim N_2((\mu_a, \mu_b), \Sigma)$,
   iii. $\gamma_{id(j)} \sim N(0, \sigma^2_{d(j)})$,

where $N_2(x, y)$ denotes a bivariate normal distribution with mean x and covariance matrix y. The Bayesian TRT model specification was completed by using a slightly informative normal hyperprior on the vector of means, inverse-gamma prior on the testlet variances, and inverse-Wishart hyperprior on Σ, respectively, to ensure proper posteriors.

The Bayesian aspect of the model, of special note, is that via the priors given above, there is sharing of information across items where smaller values of $\Sigma_{11} = \sigma_a^2$ and $\Sigma_{22} = \sigma_b^2$ would cause greater shrinkage of $\log(a_j)$ and b_j to the aggregate population values μ_a and μ_b, respectively; this holds similarly for person–testlet responses $\gamma_{id(j)}$ and $\sigma_{d(j)}^2$. This shrinkage is important as the amount of data for a given person–testlet combination is small, constrained as it is by the number of items in testlet $d(j)$.

14.2.2. Measuring DIF

As we foreshadowed in Chapter 12, the amount of DIF is measured by estimating a studied item's parameters separately in each of two examinee groups, using the rest of the test as the anchor in a method analogous to the likelihood ratio approach originally suggested by Lord (1980) and explicitly described by Thissen, Steinberg, and Wainer (1988) and summarized in Chapter 6. Specifically, we construct a test that separates the studied item (or studied testlet) from the rest of the test. It uses the balance of the test as $J - 1$ anchor items (or anchor testlets), and the studied item is included twice (thus, making the input file appear to have $J + 1$ items). The first time it appears (say Item 1), we include the responses from the focal group and act as though the reference group omitted the item. When it appears the second time (Item 1*), we include the reference group's responses and act as though the focal group omitted it. In both cases, it is then appropriately assumed that the data are either missing completely at random (MCAR) or missing at random (MAR) (Chapter 12). This design is shown pictorially in Figure 14.1 (an "X" indicates that the indicated group has been administered by the item). In this way, we can use Items $2 - J$ as an anchor to estimate each examinee's proficiency. The necessity of having a stable anchor means that we cannot study DIF simultaneously for all items at once (i.e., create a test of $2 \times J$ items with half the data missing), as that would lead to either a study of "impact" (not DIF) or a study of whether two groups are simply showing a difference in the mean ability.

When this is done, SCORIGHT (or whatever other Bayesian software used) provides the posterior distribution of all estimated parameters. The primary parameters of interest here are the item difficulties for Item 1 and Item 1*.

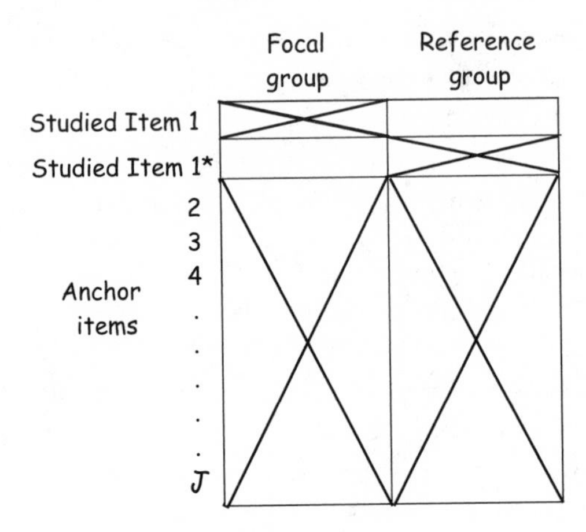

Figure 14.1. A schematic representation of the data setup to allow us to get comparable item parameters for the studied item within the reference and focal groups.

To examine these for differential functioning, we can merely take the sampled values of b_1 and b_{1^*} on each iteration beyond convergence from their posteriors and subtract them. We do this, say, 10,000 times, and look at the resulting distribution, perhaps counting the proportion of times this difference is greater than 0 (a Bayesian p-value or tail area). Of course, we might also construct the posterior distributions of both b_1 and b_{1^*} and examine the extent of their overlap as well as the other characteristics of their distributions.

14.3. The data

The data we use to illustrate our Bayesian DIF methodology are the same as those we used in Chapter 11 to illustrate the use of covariates. Specifically, they were drawn from the multiple-choice component of Step 3 of the United States Medical Licensing Examination (USMLE™). It is the third in a sequence of tests required for medical licensure, designed to assess whether a physician possesses the qualities deemed essential to assume responsibility for providing unsupervised general medical care.

Once again we focus on 54 binary items that were sampled from the multiple-choice portion of a single Step 3 form. Eleven of these were independent items and 43 were items that were nested within 19 testlets. Fourteen of these testlets were composed of two items each; the remaining five testlets were composed of three items each.

The data contained a complete set of binary scored responses for 903 examinees. Of these, 386 (43%) were female examinees and 517 (57%) were male.

14.4. A practical procedure and some results

We ran SCORIGHT with 55 items (53 anchor items plus the split DIF item as described in Section 14.2 and Figure 14.1) and 903 examinees on these data by running one long chain for 20,000 iterations ($N = 10{,}000$ iterations to converge and $M = 10{,}000$ more to provide an accurate estimate of the posterior). Previous runs of SCORIGHT, on these data and numerous others, suggested that 10,000 iterations were more than adequate for convergence. This took approximately 3 hours on a PC with a Pentium 4 processor. Clearly, doing this for every item on a test looking for DIF is impractical (or is, at least, a serious computational challenge) with current computing speeds, and is not the intent here. Rather, our interest was in demonstrating how a Bayesian-model–based approach can be used to study DIF in greater detail. In the process of this, we found that this new method also identified situations in which the MH test may struggle. In addition, if we were to run our procedure for each and every item, it would violate Paul Holland's guiding principle for practical DIF procedures ("no more than 10¢ item"). Holland's rule for DIF detection makes explicit the importance of it being cheap so as to facilitate its use, rather than giving program directors an excuse not to look for it. So, instead of examining all items using SCORIGHT, we first screened all the items using the MH statistic pioneered by Holland (Holland & Wainer, 1993). This resulted in a number of items whose MH suggested they would be worth a more careful look.[3]

For our purposes, here it is sufficient to consider just two USMLE™ items, which we shall designate Items A and B. Both items' content is psychiatry, and the MH test results (for gender DIF) are summarized in Table 14.1.

In this table, we report standard MH test results including (i) the MH chi-square statistic and its associated p value, (ii) the estimated DIF (MH D-DIF), and (iii) its associated standard error (SE (D-DIF)). Larger values of MH D-DIF, as coded here, suggest DIF effects that favor women over men. It seems clear that for both of these items, the MH statistic indicates that females significantly outperformed matched males.

[3] Finding items with significant gender DIF on this test was not a surprise. Commonly female physicians do better than males on items in the area of OB/GYN and psychiatry; males tend to do better on surgery items. This reflects differences in interest and areas of study, not bias in the items. The test is constructed to balance out this effect.

Table 14.1. *Results of an MH gender DIF analysis for two items*

Item	MH		SE (D-DIF)	P
	χ^2	D-DIF		
A	9.9	1.33	0.26	0.00
B	19.0	2.02	0.23	0.00

Table 14.2. *Item parameters estimated with the testlet model (TRT) and the usual 2-PL IRT model (BILOG)*

Item	TRT		BILOG	
	a	b	a	b
A	0.82	−1.74	0.90	−1.65
B	0.34	−5.21	0.40	−4.03

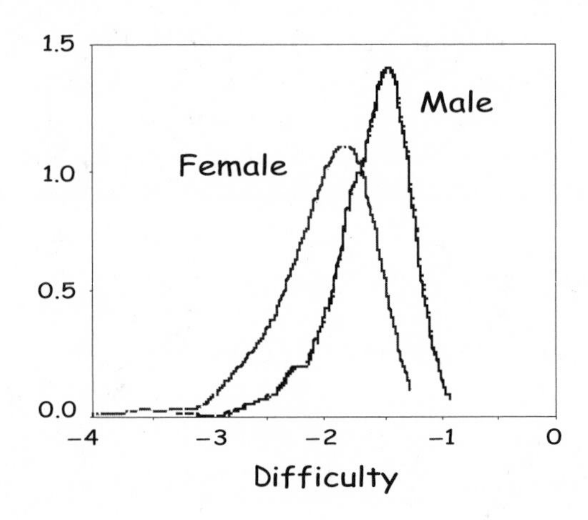

Figure 14.2. Kernel smoothed estimates of the posterior distribution of the difficulty of Item A for both male and female examinees.

To continue our "detective work," we next fit the Bayesian TRT model and found that both of these items (A and B) were very easy and not highly discriminating. We also fit the same data with BILOG (albeit not exactly the same model) to confirm these results. The resulting item parameters are in Table 14.2.

Next, we refit the TRT model as we described in Section 14.2, thus obtaining separate estimates of the parameters for Item A for male and female examinees. The posterior distributions for Item A's difficulty are shown in Figure 14.2.

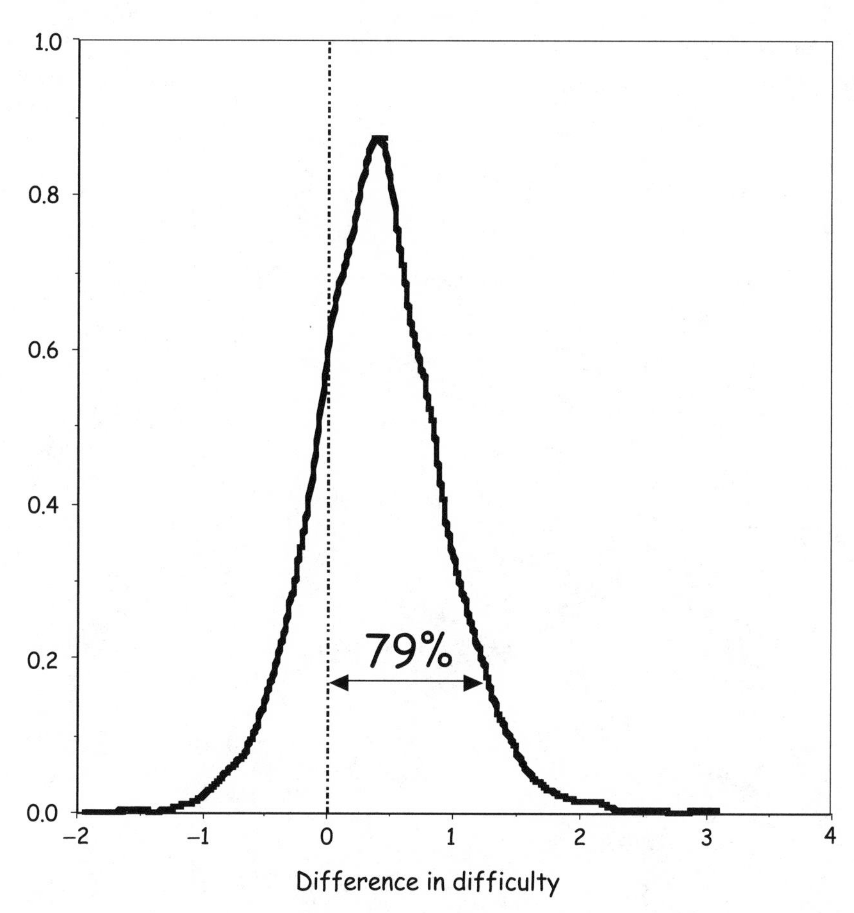

Figure 14.3. Kernel smoothed posterior density estimates of the difference between the estimates of the difficulty of Item A for male and female examinees.

It is easily seen that although this was a very easy item, male examinees found it somewhat more difficult than did female examinees. Another view of this effect was obtained by subtracting the estimate of b for female examinees from that of male examinees for 10,000 draws (after convergence) and plotting the density of that posterior distribution (Figure 14.3). We found that 79% of the time this difference was greater than 0. Perhaps a difference, but surely not as profound as might be inferred from the extreme significance level yielded by the MH analysis.

We then repeated the same analysis, but this time with Item B as the studied item. Once again (Figure 14.4) there was a small amount of evidence that male examinees found it more difficult than did female examinees, but the difference seemed very slight. Indeed, when we plotted the posterior distribution of the

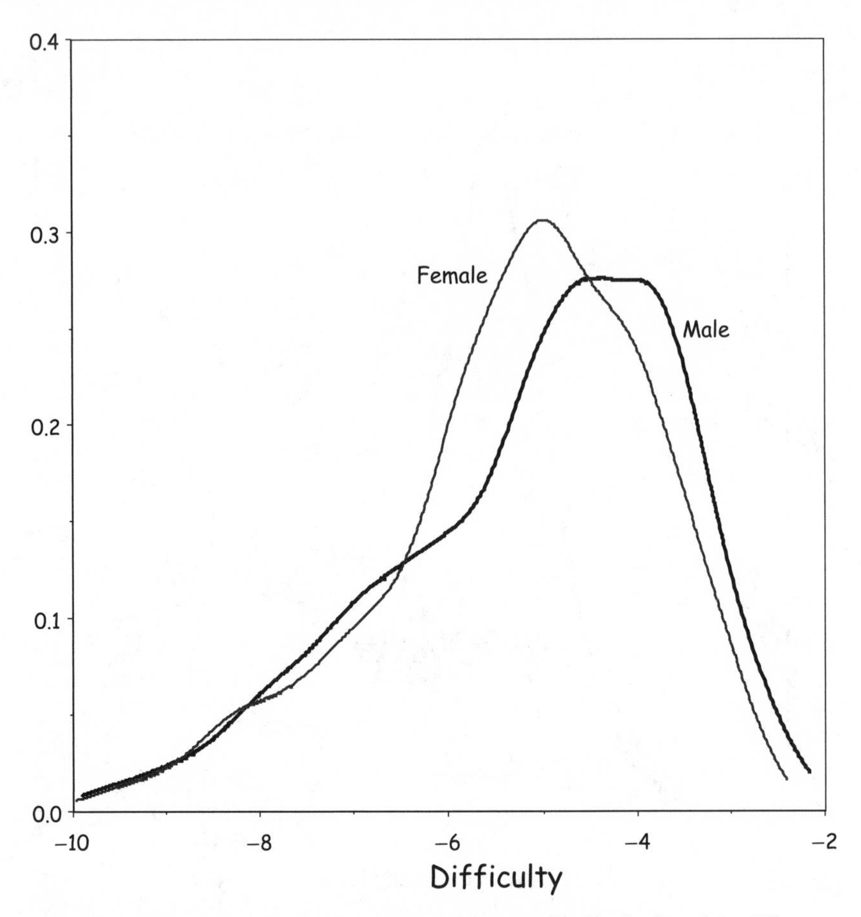

Figure 14.4. Kernel smoothed estimates of the posterior distribution of the difficulty of Item B for both male and female examinees.

difference in difficulties (Figure 14.5), we found that only 54% of the time did male examinees find Item B more difficult. This result was very different than what was portended by the MH statistic.

14.5. Simulation studies

14.5.1. Small-scale study

In view of the extensive testing that has been done in the past on the efficacy of the MH statistic in uncovering DIF, we were surprised by the difference in the inferences about these two items that would be drawn depending on the method

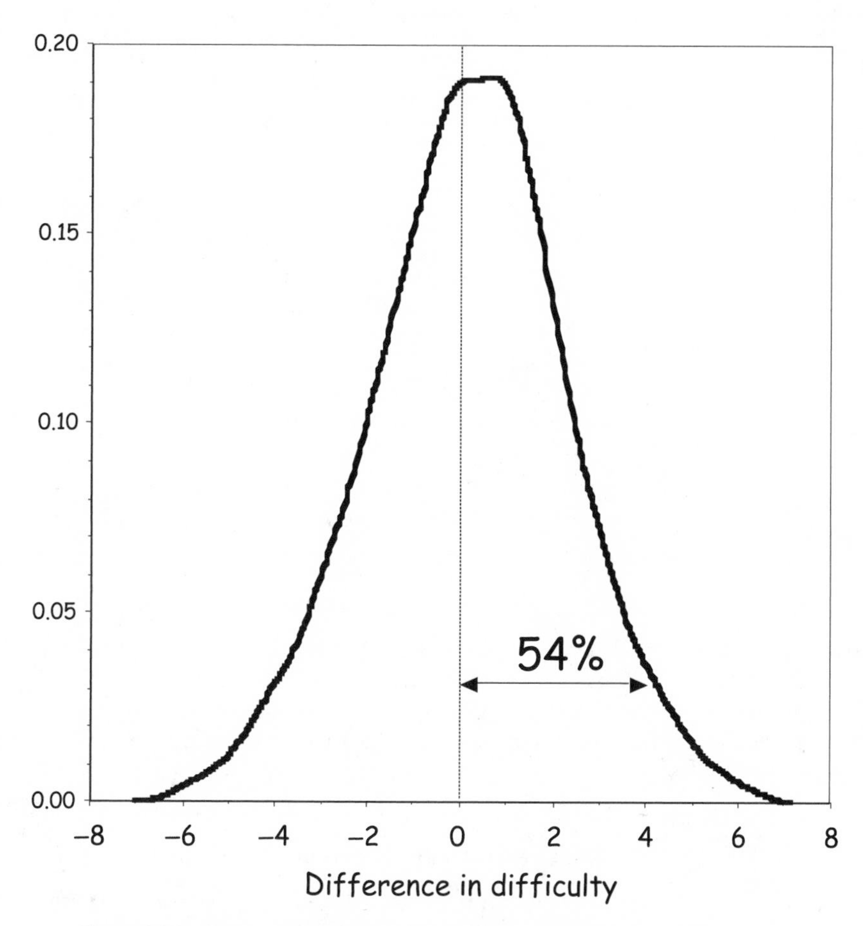

Figure 14.5. Kernel smoothed posterior density estimates of the difference between the estimates of the difficulty of Item B for male and female examinees.

of DIF detection used. Clearly, we had two different answers, but which one was right? And why? To investigate this, we did a small simulation study in which we simulated 10 data sets with the same item parameters and proficiency distribution that we had obtained from our analysis of the USMLE™ data but without any DIF. We then estimated the MH statistic and found that 15% of the simulated items showed DIF beyond the .05 level. Why did the MH's Type I error rate triple when so many prior studies showed that it performed well? We suspected that the reason is related to the values of the item discriminations (low a) and difficulties (very low b) seen here, a situation that is not unlikely in tests with a very able population (or one in which a very large fraction of examinees by design "pass"). This suspicion led us to do a much larger simulation.

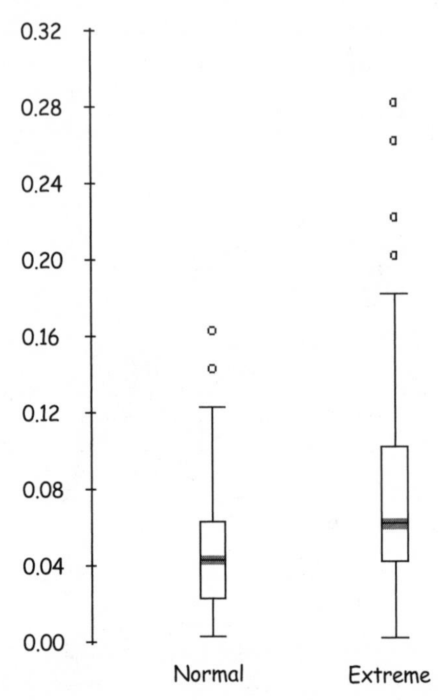

Figure 14.6. The distributions of the proportion of items the MH statistic showed with DIF at the 0.05 level for items that were of normal difficulty and for extremely easy items.

14.5.2. Large simulation study

Because the pilot study suggested that the MH statistic may be too sensitive for extremely easy and/or poorly discriminating items, we generated data using the 2-PL IRT model (TRT as given in this research would be identical) that overrepresented such items. Specifically, we simulated a 100-item test in which half the items (normal items) had $b \sim N(0, 1)$ and $a \sim N(1.25, 0.1)$ and the other half of the items (extreme items) had $b = -3.5$ and $a = 0.2$. In addition, we generated 1,000 simulees in which 500 had $\theta \sim N(0.2, 1)$ and the other 500 had $\theta \sim N(-0.2, 1)$. Thus, this data set was generated with no DIF and we wanted to assess how often it would erroneously be found. We then calculated the MH statistic for each of the two sets of items and counted how many items were found to exhibit DIF beyond the 0.05 level. We repeated this 1,000 times and the results are shown in Figure 14.6. The Type I error rate for the MH statistic ($\alpha = 0.05$) was about 4% for normal items but was 18% for extreme items. We note in passing that when we redid this simulation but set the ability distributions for the focal and reference groups equal, the difference between the normal and extreme items was not ameliorated.

We then redid the same simulations, but this time the extremely easy items had the same discriminations as the normal items ($a \sim N(1.25, 0.1)$). We recognize that this is not a realistic situation, for when an item's difficulty is well below most of the examinees' abilities, the estimates of discrimination tend to get attenuated. Nevertheless, the results were sufficiently startling to warrant mentioning. Specifically, we found that the MH statistic indicated significant DIF for 49% of these extreme items (at the 0.05 level) when the two θ distributions were equal and 45% when they were not. On the normal items, the MH coverage rate was at or near the canonical 5% level. And yet, even in these difficult circumstances, we found that the Bayesian method was not fooled; we chose 10 suspect items at random from among the extreme items and none of them was found to show significant DIF.

We suspect that the problem with the MH statistic is that the asymptotic approximation of its standard error (Phillips & Holland, 1987; Robins, Breslow, & Greenland, 1986) is too small when items are this extreme. But we leave the investigation of this issue to a later time.

14.5.3. Interpretation of the simulations

The MH test cannot find DIF on an item that is so easy that almost everyone got it right. The examinee population of STEP 3 of the USMLE™ is very able, and a high proportion of the items are answered correctly by most of the examinees. So, when we constrain the proficiency distribution to be normal with mean of 0 and standard deviation of 1, the mean difficulty of the items on the test was −3.30 (0.89) and the mean discrimination was 0.48 (0.13). There are no other studies that we know of that looked at the behavior of the MH test when the items were this extreme. Nevertheless, when we looked instead at the posterior distributions of difficulty for 10 items from these simulated "no DIF" data, we found that none of the items that the MH test had indicated had DIF showed any indication of any DIF. We had expected that the Type I error rate for the Bayesian DIF detection method we are proposing would be higher than 0.05, since the items we looked at were preselected (by the MH test) to have DIF. We were happily surprised at how well it did, and led us to expect that on a randomly selected set of items, the Bayesian method would do better still.

14.6. Conclusions

These findings bring us back to Rubin's remark quoted at the very beginning of this chapter. The MH test is widely used as the pro forma method for detecting DIF by testing companies, yet caution must be urged when the distribution of item difficulties is extreme. Yet, the Bayesian approach, likely to be far less commonly used, has a different and arguably more powerful "lens" through which to detect DIF that may be needed in certain testing circumstances (extreme items). Hence, our advice is that when one is faced with such a testing setting, it would be prudent to strongly consider using the MH test as an initial screen, but then apply a Bayesian DIF procedure as further corroboration. Availability of Bayesian software and increased computing speeds now make this a practical reality.

And to end on a possible happy note for the MH test, for this procedure seems to be conservative for extreme items, in that its Type I error rate (finding DIF when there was none) was overlarge. We did not investigate its Type II rate (not finding DIF when there is some), but leave that to subsequent accounts.

Questions

1. How is a likelihood ratio approach to DIF model testing akin to the Bayesian approach?
2. How do they differ?
3. What is the Mantel–Haenszel (MH) statistic?
4. Under what circumstances does the MH test seem to fail?
5. What is the biggest drawback of the Bayesian approach to studying DIF?
6. What are the circumstances under which the Bayesian approach is likely to be most useful?

References

Clauser, B. E., Mazor, K. M., & Hambleton, R. K. (1991). Influence of the criterion variable on the identification of differentially functioning test items using the Mantel–Haenszel statistic. *Applied Psychological Measurement, 15*, 353–359.

Clauser, B. E., Mazor, K. M., & Hambleton, R. K. (1993). The effects of purification of the matching criterion on the identification of DIF using the Mantel–Haenszel procedure. *Applied Measurement in Education, 6*, 269–279.

Clauser, B. E., Mazor, K. M., & Hambleton, R. K. (1994). The effect of score group width on the detection of differentially functioning items with the Mantel–Haenszel procedure. *Journal of Educational Measurement, 31*, 67–78.

Donoghue, J. R., Holland, P. W., & Thayer, D. T. (1993). A Monte Carlo study of factors that affect the Mantel–Haenszel and standardization measures of differential item functioning. In P. W. Holland & H. Wainer (Eds.), *Differential item functioning* (pp. 137–166). Hillsdale, NJ: Erlbaum.

Holland, P. W., & Thayer, D. T. (1988). Differential item performance and the Mantel–Haenszel procedure. In H. Wainer & H. Braun (Eds.), *Test validity* (pp. 129–146). Hillsdale, NJ: Erlbaum.

Holland, P. W., & Wainer, H. (Eds.). (1993). *Differential item functioning*. Hillsdale, NJ: Erlbaum.

Lord, F. M. (1980). *Applications of item response theory to practical testing problems.* Hillsdale, NJ: Erlbaum.

Mantel, N., & Haenszel, W. (1959). Statistical aspects of the analysis of data from retrospective studies of disease. *Journal of the National Cancer Institute, 22*, 719–748.

Mazor, K. M., Clauser, B. E., & Hambleton, R. K. (1992). The effect of sample size on the functioning of the Mantel–Haenszel statistic. *Educational and Psychological Measurement, 52*, 443–451.

Mazor, K. M., Clauser, B. E., & Hambleton, R. K. (1994). Identification of non-uniform differential item functioning using a variation of the Mantel–Haenszel statistic. *Educational and Psychological Measurement, 54*, 284–291.

Phillips, A., & Holland, P. W. (1987). Estimators of the variance of the Mantel–Haenszel log-odds-ratio estimate. *Biometrics, 43*, 425–431.

Robins, J., Breslow, N. E., & Greenland, S. (1986). Estimators of the Mantel–Haenszel variance consistent in both sparse data and large-strata limiting models. *Biometrics, 42*, 311–324.

Rogers, H. J., & Swaminathan, H. (1993). A comparison of the logistic regression and Mantel–Haenszel procedures for detecting differential item functioning. *Applied Psychological Measurement, 17*(2), 105–116.

Thissen, D., Steinberg, L., & Wainer, H. (1988). Use of item response theory in the study of group differences in trace lines. In H. Wainer & H. Braun (Eds.), *Test validity* (pp. 147–169). Hillsdale, NJ: Erlbaum.

15

A Bayesian primer

Chapters 7 to 14 of this book have made heavy use of Bayesian ideas as well as the technology of Gibbs sampling that has made those ideas practical. In this chapter, we provide a modest tutorial on these methods to help get those unfamiliar with these methods "over the hump." We do not pretend that this will provide a deep understanding for most previously naïve readers (for such an understanding we refer you to the usual sources, e.g., Gelman et al., 2003; Gilks, Richardson, & Spiegelhalter, 1995), but we do hope that this provides at least an outline of the methodology.

As part of both the good news and bad news associated with moving from true score theory–based inference (Chapter 2) to IRT-likelihood–based inference (Chapters 3–6) and finally Bayesian inference (Chapters 8–12) are the computational challenges that the researcher must face. Many of true score theory's results can be obtained by utilizing either standard analysis of variance routines (that estimates sources of variation yielding reliability estimates and the like) or factor analysis (that provides summary results on the covariation among the items); however, results under parametric IRT models require the fitting of models.

Moving beyond likelihood-based IRT to, for example, the approach taken in this book yields even greater computational challenges, as we append likelihood-based IRT models with prior distributions to yield posterior distributions as the quantities of interest, and subsequently Bayesian inference as the method of choice. And, even further, our interest is not simply on maximizing the posterior or obtaining asymptotic standard errors, that is, maximum a posteriori inference and Hessian matrix calculations (note that neither of these is more difficult for posteriors than for likelihoods); rather, our interest lies in understanding the entire posterior distribution and not just its behavior around the mode. Hence, as is common in modern Bayesian inference, our goal is to obtain samples from the posterior distributions of interest: for once obtained, this facilitates straightforward inferences that can simply be done by basic counting, averaging to compute means, sorting to compute quantiles, and so forth. It is the basic underlying aspects of Bayesian inference, and in particular as applied to IRT, and of course TRT, that we discuss here.

The remainder of this chapter, therefore, is as follows. First, we describe mathematically the general inference problem from the Bayesian point of view and why it poses significant computational challenges. Note that these are exactly the type of challenges that are met by the SCORIGHT 3.0 software (Wang, Bradlow, & Wainer, 2004) that we utilize in this book and are described throughout our examples. Second, we take the general Bayesian problem (from the previous section) and apply it to IRT models and the challenges that arise in this case. Third, we describe the computational approach – Markov chain Monte Carlo (MCMC) sampling (Gelfand & Smith, 1990) that facilitates the posterior inferences. Finally, we conclude with some advice for those doing Bayesian inference in the context of IRT models.

15.1. General Bayesian inference problem

In reviewing the general Bayesian inference problem, we provide a structure that is fairly general; however, it is primarily meant to cover the data that are typically seen in educational testing. In particular, imagine a set of I respondents on whom we have N_i measurements each. This is a classic repeated-measures data set that is quite general. For each respondent and each measurement, let us assume a scalar dependent variable given by y_{ij}, the outcome of respondent $i = 1, \ldots, I$ on measurement $j = 1, \ldots, N_i$, where $Y = (y_{ij})$ denotes the dependent variable matrix. Furthermore, let us assume that there are a set of parameters, λ_{ij}, that govern the probabilistic nature of the outcomes y_{ij} that are linked together with density function $p(y_{ij}|\lambda_{ij})$, where $\lambda = (\lambda_{ij})$ denotes the set of parameters. This function is called the likelihood function, as it provides (as described below) information on the likelihood of various parameter values (i.e., how likely is a particular value of λ_{ij} conditional on the observed data). For example, imagine a simple coin-flipping model[1] where one observes heads, $y_{ij} = 1$, with probability p_{ij}, and tails, $y_{ij} = 0$, with probability $1 - p_{ij}$. Then, the likelihood for p_{ij} conditional on observing a data point y_{ij} is given by $P(y_{ij}|\lambda_{ij} = p_{ij}) = p_{ij}^{y_{ij}}(1 - p_{ij})^{1-y_{ij}}$. From this, one can see that the more "heads" one observes, the higher is one's estimate of p_{ij}, and vice versa.

[1] Note: A binary IRT model as discussed here is a simple coin-flipping model where the probability of getting an item correct, getting "heads," is related to the ability of the person and parameters associated with the items.

As an aside from the frequentist point of view, this is the function when $p(y_{ij}|\lambda_{ij})$ is multiplied across all respondents and measurements, assuming conditional independence, yielding $p(Y|\lambda)$ that is maximized (to find the mode) and the curvature is computed (allowing computation of the standard error of the parameter estimate), to provide maximum likelihood inferences (MLEs). That is, from a frequentist perspective, one attempts to find the values of λ that maximizes the likelihood of the data that has been observed, Y.

Now, returning to the Bayesian inference problem at hand, let us assume that the parameters of the likelihood λ_{ij} are governed by a set of parameters Λ that are linked through a set of *prior* distributions $\pi(\lambda_{ij}|\Lambda)$. These are called *prior distributions* because they can be interpreted as reflecting the information about the likelihood parameters prior to observing the data, whereas $p(Y|\lambda)$ reflects the information from the data Y about the parameters λ. It is of importance to note a number of things about the prior distribution $\pi(\lambda_{ij}|\Lambda)$ from both a theoretical and practical perspective. Some of what is stated next is "historical," some of it is "our opinion," and some of it reflects areas of active (not "resolved") research.

One of the historical controversies of the Bayesian perspective is the need for a prior distribution and the impact of that prior on posterior inferences. There is concern that an improperly chosen prior can bias posterior inferences, and hence in some ways the need for a prior is seen as an additional source of uncertainty introduced into the inference process, perhaps unnecessarily. Most believe that with proper attention these risks are manageable and the benefits accrued are well worth the extra care required. We discuss a few issues surrounding the inclusion and choice of priors next.

First, the introduction of priors on the likelihood parameters is fundamentally correct in the sense that the parameters λ_{ij} are indeed unknown random quantities. Hence, by putting prior distributions on their values, we provide a coherent way to incorporate the uncertainty in their values into our posterior inferences. Second, prior distributions allow us to incorporate information into the model that may exist. This may come from prior studies, managerial judgment, exogenous data, and the like. In educational testing, there can be significant knowledge about the parameters that govern the model that should be, and in current practice are, incorporated into the Bayesian IRT models.[2] Third, by positing a prior distribution, one obtains posterior distributions that are the inferential target from a Bayesian perspective. We note that if one wants

[2] When IRT was first introduced, examinees that answered all items correctly (or all incorrectly) were unable to be given a finite score. It wasn't long before priors were introduced to provide a graceful solution. The same problem, with the same solution, arose for very easy and very difficult items.

to do inferences that are more frequentist like, the researcher is free to put uninformative priors on λ_{ij} that will yield posteriors that are proportional to the likelihood, the function that non-Bayesians also utilize for inference.

However, this does not imply that Bayesian computational methods are not of value under uninformative priors because we would still want to obtain samples from the posterior distribution, rather than estimate the mode and curvature, as finite sample inferences would still be desirable. For an example, in Chapter 13, we demonstrate the use of Bayesian methods in evaluating the probability that someone is above the cut score in a pass/fail examination (in which the probability of passing is done by simply counting the samples of the person's proficiency above the cut score) and we compare those results to maximum likelihood inferences that are derived from computing a point estimate and an asymptotic standard error. We found both finite sample benefits and differences between the approaches, demonstrating its efficacy. Hence, in our view, even when one is doing "uninformative" or "no-prior knowledge" Bayes, it still pays to be Bayesian; of course the price of eating this Bayesian omelet is the broken Bayesian egg of increased computational cost described in Chapters 8 to 10.

Last, and maybe most important, although the selection of the prior does add a source of uncertainty, (i) the likelihood can be just as wrong as the prior, hence the same skepticism and need for model checking that is directed at the prior should also be aimed at the likelihood; (ii) modern Bayesian methods allow one to assess the sensitivity of inferences to the choice of prior by running the model under differing specifications and comparing the likelihood of each model given the data using Bayes factors (Congdon, 2001; Chapter 10); and (iii) current research in the area of Bayesian semiparametric and nonparametric methods have the potential of providing a flexible class of priors (e.g., Dirichlet process priors; Ferguson, 1973), and hence less sensitivity to the underlying choices.

To complete the model specification, we choose to assume a set of distributions that govern the parameters of the prior, $\pi(\lambda_{ij}|\Lambda)$, which we denote by $\pi(\Lambda)$. This is commonly called the *hyperprior*, as it recognizes the uncertainty in the knowledge of the prior parameters. In many applications, the hyperpriors are chosen to be proper (to ensure proper posteriors) but only as slightly informative so as not to influence the posterior inferences unduly. Furthermore, in many cases, they are chosen as conjugate distributions to the priors, $\pi(\lambda_{ij}|\Lambda)$, to allow for simple Bayesian inference, as they combine easily and the resulting posterior distributions are within the same family as the assumed hyperprior, $\pi(\Lambda)$. One thing of note, and again not the main thrust of this book, if one were to simply estimate Λ and set it equal to its maximum marginal likelihood value, $\hat{\Lambda}$, this would be what is called an "empirical Bayes" approach (Morris,

1983), where one estimates the parameters of the prior from the data. We do not espouse this approach here, despite the computational advantages that it can provide, because unless one's data set is particularly large, the parameters of the prior are estimated with error and this is another source of uncertainty one may want to incorporate into the inferences.

For example, for the hyperpriors, means of distributions are commonly assumed to follow Gaussian distributions, with 0 mean and large (but not infinite) variances. Priors for variance components may be assumed to follow an inverse-gamma distribution where both the shape and scale parameter are assumed to be small, reflecting large uncertainty in their values. That is, the inverse-gamma distribution is the conjugate prior for the variance of a normal model with unknown variance. In models with proportions, one may assume that the unknown proportion follows a beta distribution with a small value for both parameters yielding an updated beta distribution for a posterior. We note that a $\beta(1, 1)$ is a uniform distribution, and in some sense an uninformative prior (actually, it is slightly informative, i.e., worth "1 prior heads and 1 prior tails"). Finally, for multivariate normal models with covariance matrices, people commonly assume an inverse-Wishart prior for the covariance matrix, yielding an inverse-Wishart posterior. For more details on the selection of priors in a variety of settings, see Gelman et al. (2003).

Thus, in summary, the Bayesian hierarchical specification is made up of three primary components:

1. $p(Y|\lambda)$ – the likelihood
2. $\pi(\lambda|\Lambda)$ – the prior
3. $\pi(\Lambda)$ – the hyperprior

Inference then entails obtaining samples from the marginal posteriors $p(\lambda|Y)$ and $p(\Lambda|Y)$, the details of which we discuss later.

15.2. Bayesian inference for IRT models

Applying the general structure from Section 15.1 to the models discussed in this book is straightforward because IRT models lend themselves naturally to a hierarchical structure. We first describe the likelihood of individual test responses that are in the form of binary outcomes. For ease of explication, and not to confuse our modeling choices with Bayesian computational issues, we do not discuss here the extension of these models to (i) allow for polytomous responses (i.e., items scored on an ordinal rating scale), (ii) incorporate testlet structures that are the main focus of this book (recognizing that many tests today include passages/"stems" with multiple items based on them), and

(iii) covariates for both the individuals (what types of persons are more or less able), items (what type of items are more or less difficult or discriminating), and testlets (what type of testlets have items that are more or less interrelated). We leave these accounts to other chapters, and our empirical data analyses that all have these features. Second, we discuss the prior distributions that are utilized to model the parameters of the IRT models. Finally, we discuss conjugate hyperpriors that are used to complete the model specification.

As described in Section 15.1, there are three main kernels of our Bayesian IRT approach: $p(Y|\lambda)$ – the likelihood; $\pi(\lambda|\Lambda)$ – the prior; and $\pi(\Lambda)$ – the hyperprior. There are many well-established likelihoods for binary test data, ranging from a one-parameter model to a two-parameter model, the 2-PL (Birnbaum, 1968), and a three-parameter model, the 3-PL (Birnbaum, 1968). Since the three-parameter model nests the simpler models, we lay out this model and describe the other two as special cases. This reprises the fuller description in Chapters 3 and 9.

The likelihood function for the 3-PL model, $P(Y|\lambda)$, can be written as follows:

$$P(Y_{ij} = 1|\lambda_{ij} = (\theta_i, a_j, b_j, c_j)) = \prod_{i=1}^{I} \prod_{j=1}^{N_i} (c_j + (1-c_j)\,\mathrm{logit}^{-1}(a_j(\theta_i - b_j))),$$

where

- The product over all persons, i, and all items, j, is justified by the assumed conditional independence across persons and items (note that a testlet structure leads to a questioning of this assumption and this is addressed in the earlier chapters).
- $\mathrm{logit}^{-1}(x) = \exp(x)/(1 + \exp(x))$, the inverse-logit: the logistic function.
- θ_i denotes the latent proficiency of examinee i. Examinees with higher proficiency have higher values of $a_j(\theta_i - b_j)$ and hence a higher probability of getting item j correct.
- a_j, b_j, c_j are item j's discrimination (slope), difficulty, and guessing parameter, respectively. That is, items with high a_j have a greater ability to discriminate between persons of low and high proficiency. Items with higher values of b_j are found to be more difficult by examinees (as a whole) while higher values of c_j (a lower asymptote commonly called a "guessing parameter") imply that persons who are infinitely unable ($\theta_i = -\infty$) still have probability c_j of getting item j correct. Note that an item with R response categories may be expected to have values of c_j around $1/R$; albeit especially attractive (or unattractive) distracters may raise (or lower) that accordingly.

The 2-PL model is derived by setting $c_j = 0$ for all items, while the 1-PL model is accomplished by setting $c_j = 0$ and $a_j = a$. One might legitimately ask that if a more general model exists, why would anyone ever fit the 2-PL or 1-PL models? and this would be a fair question. The answer is two-fold. First, the 3-PL model has known estimation challenges in that the item characteristic curve (plot of θ_i vs. $P(Y_{ij} = 1)$) is not well identified. That is, there are typically multiple combinations of (a_j, b_j, c_j) that lead to approximately the same likelihood function, and hence it is hard for the data to allow us to distinguish among them. For example, in many simulation studies, we have found that the guessing parameter can be recovered with correlation around 0.6–0.7 with the true values, whereas when a 2-PL model is used, the discrimination and difficulty parameters are "nailed" (correlation well above 0.9). Hence, for this reason, sometimes people are hesitant in using the 3-PL model. Our experience has shown that these concerns are indeed real, but the benefits typically outweigh those concerns. Indeed, people with low abilities can guess the correct answer from a finite set of choices.

The 1-PL model's use is based primarily on computational simplicity. That is, it is well known that for the 1-PL model, the number-correct score (number of items examinee i has answered correctly) and the item-number-correct score (number of examinees who get item j correct) are sufficient for estimation of θ_i and b_j, respectively. Hence, in this manner, all serious computational challenges go away, and the 1-PL model can be based on tracking simple summaries of the data; a nice feature indeed! However, it comes with a cost, a model that for many practical testing settings is unrealistic and hence fits badly, thus yielding incorrect inferences.

The likelihood function given above has unknown parameters $\lambda = (\lambda_{ij})$, which under the Bayesian approach require prior distributions. We utilize a set of prior distributions that is both flexible as well as standard in hierarchical IRT models. In particular, we model $P(\lambda|\Lambda)$ as:

$$\theta_i \sim N(0, 1) \text{ and}$$
$$(\log(a_j), b_j, \text{logit}(c_j)) \sim N_3(\mu = (\mu_a, \mu_b, \mu_c), \Sigma).^3$$

We note that the mean and variance of the distribution of θ are fixed at 0 and 1 to identify the model. Furthermore, we assume $\log(a_j)$ to be normal (as $a_j > 0$ is a typical assumption and observed), and $\text{logit}(c_j)$ to be normal as

[3] The multivariate normal prior here is our most general model and is not utilized in Chapters 8 to 10 that describe early versions of our TRT work. Chapter 11, which represents our full model with covariates, incorporates the multivariate normal prior. Other chapters utilize independent priors for $\log(a)$, b, and $\text{logit}(c)$.

$0 \leq c_j \leq 1$, hence the logit function transforms it to something on the entire real number line. The covariance matrix among the item parameters, Σ, is used to reflect that these are typically correlated; despite this, it does add to the computational challenge and some software does not implement this choice. SCORIGHT does, and is one of the strengths of its implementation.

To complete the model specification, we need to specify a hyperprior distribution, $\pi(\Lambda)$. For this, we utilize proper conjugate priors (as discussed above) where we assume

$$\mu \sim N_3(0, v \times I_3) \text{ and}$$
$$\Sigma \sim \text{Inv-Wishart}\,(S, n_0),$$

where v is chosen to be a very large number (say 100), I_3 is an identity matrix making the prior covariance matrix for μ diagonal, and n_0 is chosen to be small (e.g., 3, the dimension of Σ) to be minimally informative. This completes the Bayesian IRT model specification. We next describe the computational challenges in obtaining inferences under this model.

15.3. Computation for the Bayesian IRT model

As mentioned previously, the inferential goal in a Bayesian model is to make inferences about the set of posterior distributions of the model parameters via posterior sampling. Unfortunately, the model described in Section 15.2 does not lend itself to straightforward sampling because the product–Bernoulli (coin-flipping) form of the likelihood in $P(Y|\lambda)$ and its associated set of Gaussian priors $p(\lambda|\Lambda)$ are not conjugate. That is, the set of marginal posterior distributions, given by

$$p(\lambda|Y) \propto \int p(Y|\lambda)p(\lambda|\Lambda)\pi(\Lambda)d\Lambda,$$

are not available in closed form. Note that an analogous nonclosed form equation could be written for the marginal posteriors of the prior distribution parameters $p(\Lambda|Y)$ as well. Hence, to perform this integration, a set of computational techniques, in particular MCMC methods, have been developed that allow for sampling from posteriors of this type. We provide next a brief introduction to how it is done. Prior chapters provide additional details and illustrations.

The way in which MCMC methods work is by setting up a Markov chain (each iteration of the MCMC sampler's movement depends only on where it was on the previous iteration) that converges in distribution to the stationary distribution of interest. The way in which this is accomplished for the Bayesian IRT model is via the following algorithmic steps:

1. Let t = iteration number = 0. Obtain an initial estimate for parameters λ and Λ, denote them by $\lambda^{(0)}$ and $\Lambda^{(0)}$.
2. Let $t = t + 1$. Draw a sample from the conditional distribution $p(\lambda^{(t)}|\Lambda^{(t-1)}, Y)$.
3. Draw a sample from the conditional distribution $p(\Lambda^{(t)}|\lambda^{(t)}, Y)$.
4. If $t \leq M$, then go to [2], where $M' < M$ is the number of iterations to burn in and $M - M'$ is the number of iterations used for estimation.

A description of this process is as follows. In Step 1, one utilizes an initial guess of the parameters to start the Markov chain sampler. Much research with IRT models suggests that convergence to the stationary distribution (the desired result) is not overly sensitive to the starting values. Typically, starting values are obtained by either generating values from the prior, that is, $\theta_i \sim N(0, 1)$, the so-called random start procedure, or utilizing values obtained from exogenous methods. For instance, one could run a mode-finding algorithm (e.g., EM algorithm – Dempster, Laird, & Rubin, 1977) and utilize it as a starting value. This will not only provide greater assurance of convergence of the Markov chain but also lead to an increased speed of convergence, as the algorithm is starting in approximately the correct place. Alternatively, one could obtain starting values from a "canned" software package that is finding maximum likelihood estimates from approximately the same model, for example, BILOG (Mislevy & Bock, 1983), and input those as starting values. SCORIGHT, as described later, allows for both of these starting procedures.

In Step 2, one has to obtain a sample from the conditional distribution $p(\lambda^{(t)}|\Lambda^{(t-1)}, Y)$, where $\lambda^{(t)} = (\theta_i, a_j, b_j, c_j)$, the parameters that govern the IRT model. Although this goal is easy to "state" (as in Step 2), it is not trivial to implement because

$$p(\lambda^{(t)}|\Lambda^{(t-1)}, Y) \propto p\left(Y|\lambda^{(t)}, \Lambda^{(t-1)}\right) p\left(\lambda^{(t)}|\Lambda^{(t-1)}\right).$$

Moreover, it is not a standard model because $p(Y|\lambda^{(t)}, \Lambda^{(t-1)})$ and $p(\lambda^{(t)}|\Lambda^{(t-1)})$ are not conjugate as the former is an IRT model and the latter, per Section 15.2 is a set of Gaussian priors.

There has been much research over the past 15 or so years that enables sampling from conditional distributions such as $p(\lambda^{(t)}|\Lambda^{(t-1)}, Y)$. We discuss the most popular one here, and it is what that is utilized in SCORIGHT, the Metropolis–Hastings algorithm (Chib & Greenberg, 1995). The way in which the Metropolis–Hastings algorithm works is that if one wants to obtain a sample from a *target density*, say $f(\theta)$, which does not admit straightforward sampling, one proceeds as follows.[4]

i. Choose a *sampling* density $g(\theta)$ that does admit straightforward sampling.
ii. Obtain a sampled value from $g(\theta)$; denote this value as θ^*.
iii. Compute $f(\theta^*)$ and $g(\theta^*)$, the height of the target and sampling density at θ^*.
iv. Compute $r = \min(1, f(\theta^*)/g(\theta^*))$, and accept the drawn value of θ^* with probability r (i.e., $\theta^{(t+1)} = \theta^*$); otherwise, let $\theta^{(t+1)} = \theta^{(t)}$.

Per Step iv, the Metropolis–Hastings algorithm either keeps the MCMC sampler at its current value from the previous iteration, $\theta^{(t)}$, or accepts the drawn value θ^*. One somewhat open issue today is how to choose the sampling density g(θ) and what the target acceptance rate should be. While one might assume that one should choose g(θ) so that r is near 1, recent research (Gelman, Gilks, & Roberts, 1995) actually shows this not to be the case. They suggest that an acceptance rate between 20% and 40% may be optimal. The rationale for their argument is that if one makes the acceptance rate too high, then the MCMC sampler moves "too slowly" through the parameter space, that is, it is takes too small a jump at each iteration. Alternatively, if one was to reject some draws, θ^*, then when one moves from $\theta^{(t)}$ to θ^*, the jumps will tend to be larger, giving the MCMC sampler better mixing properties.

One popular way to select a jumping kernel is a random-walk Metropolis, where one chooses $g(\theta) \sim N(\theta^t, c \times \sigma^2)$, a normal distribution centered around the previous draw, $\theta^{(t)}$, with variance $c \times \sigma^2$, where c is chosen to get the desired acceptance rate of 20% to 40% as described earlier. Running a random-walk Metropolis is relatively easy and typically leads to good mixing.

In Step 3 of the MCMC algorithm, that is sampling from $p(\Lambda^{(t)}|\lambda^{(t)}, Y)$, is straightforward, as described in Section 15.1, $\pi(\Lambda)$ is typically chosen to be conjugate to $p(\lambda|\Lambda)$ and from a familiar sampling density. Hence, this conditional distribution can be sampled directly in the way most Bayesian IRT models are specified.

Finally, Step 4 of the MCMC algorithm states to run the sampler for M iterations, where M' iterations are those that are utilized to burn-in the sampler, that is, move from the initial starting values (Step 1) to a time in which draws are being obtained from the stationary distribution. Then, after convergence is obtained, one samples $M - M'$ additional draws to be used for inference (e.g.,

[4] The Metropolis–Hastings algorithm described here assumes a symmetric sampling density or "jumping kernel." If it is not, there is an additional term in Step iv that relates to the probability of jumping under an asymmetric jumping distribution. We leave that for more detailed descriptions of the algorithm.

computing means, variances, quantiles, confidence intervals). An issue on this topic that is worth discussing, and is currently implemented in SCORIGHT, is how to determine M', the number of iterations to utilize as burn-in. On the basis of a method developed by Gelman and Rubin (1992), one typically proceeds by running multiple (say 3–5) independent MCMC chains (from overdispersed starting values) and then by diagnosing convergence of the multiple chains (to the same stationary distribution) using an F-test that is provided. The F-test provides a test of the across-chain variability to within-chain variability. When this ratio is low, this suggests convergence. In addition, for each parameter in the MCMC chain, the method provides an interval called "confshrink," which is how much the parameter's posterior uncertainty would shrink if the Markov chain were to be continued to run "forever." Gelman and Rubin demonstrated that if this value is below 1.2 for each parameter in the model, there has likely been adequate convergence. Our experience with Bayesian IRT models, and with SCORIGHT, suggests that they are well-behaved models and ones in which convergence could be achieved (of course, less so for the 3-PL model as compared to the 2-PL model or 1-PL model). For a more complete description of convergence issues for Bayesian IRT models, see Sinharay (2004).

15.4. Some advice for those using Bayesian IRT models

As we conclude this chapter, we provide some advice on the basis of our thirty plus years of combined experience running Bayesian IRT models. These pieces of advice come in no particular order; rather, we hope that users adhere to all that apply to their given testing situation.

1. Let the data and testing situation under consideration drive the appropriate model choice. That is, while there are computational advantages to running a 1-PL or a non-Bayesian model, these decisions should be based on verifying the assumptions that underlie the simpler model. Modern *free* software, whether it is SCORIGHT, WinBUGS (http://www.mrc-bsu.cam.ac.uk/bugs/), or similar programs have made computing much more feasible; we hope to the degree that researchers will choose to implement them.
2. When running a Bayesian IRT model, simulate data from the model to verify the computational accuracy of the approach used. That is, as in Bradlow, Wainer, and Wang (1999) and given in Chapter 8, it is important to confirm that the MCMC sampler is properly sampling from the stationary distribution of interest. Our experience is that one would learn

very quickly whether proper computation is being done, as the draws obtained would be far from the truth if errors are occurring.

3. When applying Bayesian IRT methods to real data, run a more standard software package such as BILOG first to both obtain reasonable starting values and also to provide a set of benchmark results to which one can compare. If BILOG says one thing and the Bayesian IRT model says something radically different, then this is an area for careful investigation.
4. Run multiple chains from overdispersed starting values and test for convergence. Of course, this does increase computation time (typically linearly in the number of chains run); however, this provides an important degree of model checking. This can be implemented, as discussed, by sampling from the prior distribution (i.e., sampling $\theta \sim N(0, 1)$).[5]

This simplified summary of Bayesian estimation for the most basic of IRT models was meant to provide a boost to the use of this new technology for those in need of it. The prior chapters provide generalizations of this to testlets, to polytomous models, and to the use of covariates within the model.

Questions

1. Explain why draws obtained from a Markov chain Monte Carlo sampler are serially correlated and the implications that has for obtaining standard errors of parameter estimates.
2. To overcome serial autocorrelation, as described in Question (1), people sometimes "thin" the output from the Markov chain taking every k-th, $k \geq 1$, draw instead after convergence.
   i. Describe why how thinning handles the serial autocorrelation problem.
   ii. Describe how one might determine k, the thinning interval.
   iii. Describe why thinning makes more sense from a frequentist perspective than from a Bayesian perspective. Hint, what is the inferential quantity of interest under each perspective.
3. Describe the difference between utilizing parametric, semiparametric, and nonparametric priors? What are the advantages and disadvantages of each?
4. Describe why empirical Bayes inferences may lead to posterior intervals that are too narrow and how bootstrapping may be an approach to ameliorate some of the issues.
5. Explain the advantage of using a multivariate normal prior for the item parameters as opposed to independent priors? In what testing settings would you imagine greater or lower correlation among the parameters?

[5] To assess whether the prior distribution provides adequate overdispersion, if there is concern, one can run a few chains by sampling from the prior and see the path of the MCMC chains.

References

Birnbaum, A. (1968). Some latent trait models and their use in inferring an examinee's ability. In F. M. Lord & M. R. Novick (Eds.), *Statistical theories of mental test scores* (pp. 392–479). Reading, MA: Addison-Wesley.

Bradlow, E. T., Wainer, H., & Wang, X. (1999). A Bayesian random effects model for testlets. *Psychometrika, 64*, 153–168.

Chib, S., & Greenberg, E. (1995). Understanding the Metropolis–Hastings algorithm. *The American Statistician, 49*, 327–335.

Congdon, P. (2001). *Bayesian statistical modeling* (pp. 465–493). England: Wiley.

Dempster, A. P., Laird, N. M., & Rubin, D. B. (1977). Maximum likelihood from incomplete data via the EM algorithm (with discussion). *Journal of the Royal Statistical Society, Series B, 39*, 1–38.

Ferguson, T. S. (1973, March). A Bayesian analysis of some nonparametric problems. *The Annals of Statistics, 1*(2), 209–230.

Gelfand, A. E., & Smith, A. F. M. (1990). Sampling-based approaches to calculating marginal densities. *Journal of the American Statistical Association, 85*, 398–409.

Gelman, A., Carlin, J. B., Stern, H. S., & Rubin, D. B. (2003). *Bayesian data analysis* (2nd ed.). New York: Chapman & Hall.

Gelman, A., Gilks, W. & Roberts, G., (1995). Efficient Metropolis jumping rules. In *Bayesian Statistics 5*. New York: Oxford University Press.

Gelman, A., & Rubin, D. B. (1992). Inference from iterative simulation using multiple sequences. *Statistical Science, 7*, 457–472.

Gilks, W. R., Richardson, S., & Spiegelhalter, D. J. (1995). *Markov chain Monte Carlo in practice*. New York: Chapman & Hall.

Mislevy, R. J., & Bock, R. D. (1983). BILOG: Item analysis and test scoring with binary logistic models [Computer program]. Mooresville, IN: Scientific Software.

Morris, C. N. (1983). Parametric empirical Bayes inference: Theory and applications (c/r: P55–65). *Journal of the American Statistical Association, 78*, 47–55.

Rasch, G. (1960). *Probabilistic models for some intelligence and attainment tests.* Copenhagen: Denmarks Paedagogiske Institut.

Sinharay, S. (2004). Experiences with Markov chain Monte Carlo convergence assessment in two psychometric examples. *Journal of Educational and Behavioral Statistics, 29*, 461–488.

Wang, X., Bradlow, E. T., & Wainer, H. (2004). A user's guide for SCORIGHT (Version 3.0): A computer program built for scoring tests built of testlets including a module for covariate analysis. (ETS Research Report RR 04-49). Princeton, NJ: Educational Testing Service.

Wishart, J. (1928). The generalized product moment distribution in samples from a normal multivariate population. *Biometrika, 20A* (32), 424.

Glossary of terms

Bayes' factor. In Bayesian statistics, two rival models may be compared on the basis of the ratio of marginal likelihoods, known as the Bayes' factor. It is a multiple of the usual likelihood ratio where the ratio involves the prior distributions of the two rival models.

Beta distribution. This term is usually applied to the form

$$dF = \{[x^{\alpha-1}(1-x)^{\beta-1}]/B(\alpha, \beta)\}dx, 0 \leq x \leq 1; \alpha, \beta > 0.$$

Conjugate prior. A prior distribution for a parameter for which there is a sufficient statistic is conjugate if the prior and posterior distributions have the same functional form. An example is the beta prior for the probability in a binomial distribution.

Dirichlet process priors. The Dirichlet process prior allows more flexible nonparametric mixture modeling than finite mixture models that assume a certain number of mixture classes and a particular parametric form (e.g., a K-variate normal mixture). In a Dirichlet process prior, the number of mixture components is not specified in advance and can grow as new data come in.

Informative prior. A prior probability density function that provides positive empirical or theoretical information regarding the value of an unknown parameter. The contrary case, where little or no information about the parameter is available, is represented by a diffuse or vague prior.

Gamma distribution. A frequency distribution of the form

$$dF = \{[e^{-\xi}x^{\lambda-1}]/\Gamma(\lambda)\}dx, 0 \leq x \leq \infty.$$

It is also known as Pearson's Type III or simply Type III distribution.

Hessian matrix. A Hessian matrix, H, is a matrix of second-order partial derivatives, typically of the likelihood function. When one computes $-H^{-1}$, its negative inverse, this yields a matrix with diagonal elements that are the asymptotic variances for the model parameters when it is evaluated at the MLE and was called the "Information matrix" by R. A. Fisher.

Hyperprior. The hyperprior is the set of distributions that are put on the parameters that govern the prior distribution. For example, if we define the likelihood as $p(Y|\lambda)$, the prior distribution as $p(\lambda|\Lambda)$, then $p(\Lambda)$, a distribution on the parameters of the prior, is called the hyperprior.

Inverse-gamma. The inverse-gamma distribution has the probability density function over the support $x > 0$

$$f(x; \alpha, \beta) = \frac{\beta^{\alpha}}{\Gamma(\alpha)} x^{-\alpha-1} \exp\left(\frac{-\beta}{x}\right)$$

with shape parameter α and scale parameter β. It is the conjugate prior distribution for the variance component that is used in many hierarchical models.

Inverse-Wishart. The inverse-Wishart distribution is the multivariate generalization of the inverse-gamma distribution and is the conjugate prior distribution for covariance matrix of a multivariate normal distribution.

Improper prior. A probability distribution with a distribution function $F(x)$ for which $F(-\infty) > 0$ or $F(+\infty) < 1$.

Jumping kernel. When utilizing the Metropolis–Hastings algorithm, one generates candidate values for the next iteration, $t + 1$, say θ^*, for which one is considering moving to from $\theta^{(t)}$. The jumping kernel is the distribution from which θ^* is sampled from and is chosen to be an easy-to-sample-from distribution. See the definition for the Metropolis–Hastings algorithm as to the probability with which one moves from $\theta^{(t)}$ to θ^*.

Marginal posteriors. Inferences for model parameters are desired that are unconditional on the values of other parameters. When one obtains the marginal posterior distribution, one does this by integrating out the other parameters of the model, allowing unconditional inferences for the parameter of interest conditional only on the observed data.

Maximum a posteriori (MAP) inference. The posterior distribution is proportional to the likelihood x prior. Finding the most likely values of the parameters of the posterior distribution, conditional on the observed data is called MAP inference. It is conceptually identical to maximum likelihood estimate that maximizes the likelihood (i.e., no prior).

Metropolis–Hastings. The Metropolis–Hastings algorithm is a method by which one can obtain samples from a target density g that is not amenable to closed-form sampling by obtaining samples from a sampling density f and accepting those draws with probability $r = \min(1, f/g)$.

Proper prior. A proper probability distribution is one with a distribution function $F(x)$ for which $F(-\infty) = 0$ and $F(+\infty) = 1$. A proper prior distribution is used to reflect uncertainty in the parameters of the likelihood and/or incorporate prior knowledge, while ensuring that the posterior distribution will be proper.

Stationary distribution. A stochastic process $\{x_t\}$ is stationary if the multivariate distribution of $x_{t(1+h),} x_{t(2+h),...,} x_{t(n+h)}$ is independent of h for any finite set of parameter values. The process is said to be stationary in the wide sense if the mean and variance exist and are independent of t. Thus, when one runs a Markov chain, reaching the stationary distribution implies that one is sampling from the distribution of interest and that the effect of the starting values of the algorithm have been essentially removed.

Wishart distribution. The joint distribution of variances and covariances in samples from a p-variate normal population, derived by Wishart (1928). It is commonly utilized as a prior distribution for the inverse of a covariance matrix because it is conjugate.

Epilogue

Jephthah then called together the men of Gilead and fought against Ephraim. The Gileadites struck them down because the Ephraimites had said, "You Gileadites are renegades from Ephraim and Manasseh."

The Gileadites captured the fords of the Jordan leading to Ephraim, and whenever a survivor of Ephraim said, "Let me cross over," the men of Gilead asked him, "Are you an Ephraimite?" If he replied, "No," they said, "All right, say 'Shibboleth.'[1] "He said, "Sibboleth," because he could not pronounce the word correctly, they seized him and killed him at the fords of the Jordan.

Forty-two thousand Ephraimites were killed at that time.
Judges 12:4–6

The bible records an early example of testing in a one-item test that was administered. It had very high stakes, and no validity or reliability studies were reported in support of its conclusions. The author of Judges did not record what proportion of those slain were, in fact, Ephraimites, and how many were unfortunate Gileadites with a speech impediment. Happily the science of testing has advanced considerably in the three thousand odd years since Jephthah's struggle with Ephraim.[2]

[1] **Shibboleth** (ʃɪbəlɛθ) is any language usage indicative of one's social or regional origin, or more broadly, any practice that identifies members of a group. It comes from the Hebrew word (שבולת) that literally means "ear of grain" or "torrent of water." In the Hebrew Bible, pronunciation of this word was used to distinguish members of a group (like Ephraim) whose dialect lacked a /ʃ/ sound (as in ***shoe***) from members of a group (like Gilead) whose dialect included such a sound.

[2] The bible also contains one of the earliest examples of testing for selection, in which Lord played an important role in devising a test to whittle down 22,000 men to the 300 who were needed. (Judges 7:4–7)

But the LORD said to Gideon, "There are still too many men. Take them down to the water, and I will sift them for you there. If I say, 'This one shall go with you,' he shall go; but if I say, 'This one shall not go with you,' he shall not go."

So Gideon took the men down to the water. There the LORD told him, "Separate those who lap

Yet, we can see that the idea of a test with associated consequences is a very old one indeed. In the more than three millennia since it is not surprising that there have been modifications in, and improvements to, the practice of testing, the theory of testing, and the ethics of testing. But the soul of the idea has remained the same. In this monograph, we have tried to inch this ancient art a wee bit further forward.

the water with their tongues like a dog from those who kneel down to drink." Three hundred lapped the water with their tongues. All the rest knelt and drank from their cupped hands. The LORD said to Gideon, "With the three hundred men that lapped I will save you and give the Midianites into your hands. Let all the other men go, each to his own place."

Bibliography

Albert, J. H. (1992). Bayesian estimation of normal ogive item response curves using Gibbs sampling. *Journal of Educational Statistics, 17*, 251–269.

Albert, J. H., & Chib, S. (1993). Bayesian analysis of binary and polychotomous response data. *Journal of the American Statistical Association, 88*, 669–679.

American Psychological Association. (1966). *Standards for educational and psychological tests and manuals*. Washington, DC: Author.

American Psychological Association. (1974). *Standards for educational and psychological tests*. Washington, DC: Author.

American Psychological Association. (1985). *Standards for educational and psychological testing*. Washington, DC: Author.

Anastasi, A. (1961). *Psychological testing* (2nd ed.). New York: Macmillan.

Anastasi, A. (1976). *Psychological testing* (4th ed.). New York: Macmillan.

Andersen, E. B. (1970). Asymptotic properties of conditional maximum likelihood estimators. *Journal of the Royal Statistical Society, Series B, 32*, 283–301.

Anderson, T. W. (1971). *The statistical analysis of time series*. New York: Wiley.

Angoff, W. H. (1971). Scales, norms and equivalent scores. In R. L. Thorndike (Ed.), *Educational measurement* (pp. 508–600). Washington, DC: American Council on Education.

Angoff, W. H. (1982). Use of difficulty and discrimination indices for detecting item bias. In R. A. Berk (Ed.), *Handbook of methods for detecting item bias* (pp. 96–116) Baltimore, MD: Johns Hopkins University Press.

Angoff, W. H. (1993). Perspectives on the theory and application of differential item functioning methodology. In P. W. Holland & H. Wainer (Eds.), *Differential item functioning* (pp. 3–23). Hillsdale, NJ: Erlbaum.

Bell, E. T. (1937). *Men of mathematics*. New York: Simon & Schuster.

Bentler, P. M., & Bonett, D. G. (1980). Significance tests and goodness of fit in the analysis of covariance structures. *Psychological Bulletin, 88*, 588–606.

Birnbaum, A. (1968). Some latent trait models and their use in inferring an examinee's ability. In F. M. Lord & M. R. Novick (Eds.), *Statistical theories of mental test scores* (pp. 392–479). Reading, MA: Addison-Wesley.

Bishop, Y. M. M., Fienberg, S. E., & Holland, P. W. (1975). *Discrete multivariate analysis*. Cambridge, MA: MIT Press.

Bock, R. D. (1967, March). *Fitting a response model for n dichotomous items*. Paper presented at the annual meeting of the Psychometric Society, Madison, WI.

Bock, R. D. (1972). Estimating item parameters and latent ability when responses are scored in two or more latent categories. *Psychometrika, 37*, 29–51.

Bock, R. D. (1993). Different DIFs: Comment on the papers read by Neil Dorans and David Thissen. In P. W. Holland & H. Wainer (Eds.), *Differential item functioning* (pp. 115–122). Hillsdale, NJ: Erlbaum.

Bock, R. D., & Aitkin, M. (1981). Marginal maximum likelihood estimation of item parameters: An application of an EM algorithm. *Psychometrika, 46*, 443–459.

Bock, R. D., Gibbons, R., & Muraki, E. (1988). Full information item factor analysis. *Applied Psychological Measurement, 12*, 261–280.

Bock, R. D., & Kolakowski, R. (1973). Further evidence of sex-linked major-gene influence on human spatial visualizing ability. *American Journal of Human Genetics, 25*, 1–14.

Bock, R. D., & Lieberman, M. (1970). Fitting a response model for *n* dichotomously scored items. *Psychometrika, 35*, 179–197.

Boulet, J. R., Ben David, M. F., Ziv, A., Burdick, W. P., Curtis, M., Peitzman, S., & Gary, N. C. (1998). Using standardized patients to assess the interpersonal skills of physicians. *Academic Medicine, 73*, S94–S96.

Bradlow, E. T., & Wainer, H. (1998). Some statistical and logical considerations when rescoring tests. *Statistica Sinica, 8*, 713–728.

Bradlow, E. T., Wainer, H., & Wang, X. (1999). A Bayesian random effects model for testlets. *Psychometrika, 64*, 153–168.

Brown, W. (1910). Some experimental results in the correlation of mental abilities. *British Journal of Psychology, 3*, 296–322.

Brown, W., & Thomson, G. H. (1925). *The essentials of mental measurement* (3rd ed.). London: Cambridge University Press.

Bush, R. R., & Mosteller, F. (1955). *Stochastic models for learning*. New York: Wiley.

Chang, H.-H., & Mazzeo, J. (1994). The unique correspondence of item response functions and item category response function in polytomously scored item response models. *Psychometrika, 59*, 391–404.

Chib, S., & Greenberg, E. (1995). Understanding the Metropolis–Hastings algorithm. *The American Statistician, 49*, 327–335.

Clauser, B. E., Mazor, K. M., & Hambleton, R. K. (1991). Influence of the criterion variable on the identification of differentially functioning test items using the Mantel–Haenszel statistic. *Applied Psychological Measurement, 15*, 353–359.

Clauser, B. E., Mazor, K. M., & Hambleton, R. K. (1993). The effects of purification of the matching criterion on the identification of DIF using the Mantel–Haenszel procedure. *Applied Measurement in Education, 6*, 269–279.

Clauser, B. E., Mazor, K. M., & Hambleton, R. K. (1994). The effect of score group width on the detection of differentially functioning items with the Mantel–Haenszel procedure. *Journal of Educational Measurement, 31*, 67–78.

Cohen, J. (1960). A coefficient of agreement for nominal scales. *Educational and Psychological Measurement, 20*, 37–46.

Congdon, P. (2001). *Bayesian statistical modeling* (pp. 465–493). England: Wiley.

Cronbach, L. J. (1951). Coefficient alpha and the internal structure of tests. *Psychometrika, 16*, 297–334.

Cronbach, L. J., Gleser, G. C., Nanda, H., & Rajaratnam, N. (1972). *The dependability of behavioral measurements: Theory of generalizability for scores and profiles*. New York: Wiley.

Crumm, W. L. (1923). Note on the reliability of a test, with a special reference to the examinations set by the College Entrance Board. *The American Mathematical Monthly, 30*(6), 296–301.

Damien, P., Wakefield, J. C., & Walker, S. G. (1999). Gibbs sampling for Bayesian non-conjugate and hierarchical models by using auxiliary variables. *Journal of the Royal Statistical Society, B Statistical Methodology, 61*, 331–344.

Dempster, A. P., Laird, N. M., & Rubin, D. B. (1977). Maximum likelihood from incomplete data via the EM algorithm (with discussion). *Journal of the Royal Statistical Society, shape Series B, 39*, 1–38.

Donlon, T. F., & Angoff, W. H. (1971). The Scholastic Aptitude Test. In W. H. Angoff (Ed.), *The College Board Admissions Testing Program* (pp. 15–47). New York: College Entrance Examination Board.

Donoghue, J. R., Holland, P. W., & Thayer, D. T. (1993). A Monte Carlo study of factors that affect the Mantel–Haenszel and standardization measures of differential item functioning. In P. W. Holland & H. Wainer (Eds.), *Differential item functioning* (pp. 137–166). Hillsdale, NJ: Erlbaum.

Dorans, N. J., & Holland, P. W. (1993). DIF detection and description: Mantel–Haenszel and standardization. In P. W. Holland & H. Wainer (Eds.), *Differential item functioning* (chap. 3, pp. 35–66). Hillsdale, NJ: Erlbaum.

Duhachek, A., Coughlan, A. T., & Iacobucci, D. (2005). Results on the Standard Error of the Coefficient Alpha Index of Reliability. *Marketing Science, 24*(2), 294–301.

Efron, B., & Morris, C. (1973). Stein's estimation rule and its competitors – An empirical Bayes approach. *Journal of the American Statistical Association, 68*, 117–130.

Efron, B., & Morris, C. (1975). Data analysis using Stein's estimator and its generalizations. *Journal of the American Statistical Association, 70*, 311–319.

Eignor, D. R., & Cook, L. L. (1983, April). *An investigation of the feasibility of using item response theory in the preequating of aptitude tests.* Paper presented at the annual meeting of the American Educational Research Association, Montreal, Canada.

Ferguson, T. S. (1973, March), A Bayesian analysis of some nonparametric problems. *The Annals of Statistics, 1*(2), 209–230.

Fisher, R. A. (1922). On the mathematical foundations of theoretical statistics. *Philosophical Transactions of the Royal Society of London A, 222*, 309–368.

Fisher, R. A. (1925). Theory of statistical estimation. *Proceedings of the Cambridge Philosophic Society, 22*, 700–725.

Gelfand, A. E., & Smith, A. F. M. (1990). Sampling-based approaches to calculating marginal densities. *Journal of the American Statistical Association, 85*, 398–409.

Gelman, A., Carlin, J. B., Stern, H. S., & Rubin, D. B. (1995). *Bayesian data analysis.* London: Chapman & Hall.

Gelman, A., Carlin, J. B., Stern, H.S., & Rubin, D. B. (2003). Bayesian data analysis (2nd ed.). New York: Chapman & Hall.

Gelman, A., Roberts, G., & Gilks, W. (1995). Efficient Metropolis jumping rules. In *Bayesian statistics 5.* New York: Oxford University Press.

Gelman, A., & Rubin, D. B. (1992). Inference from iterative simulation using multiple sequences. *Statistical Science, 7*, 457–472.

Geman, S., & Geman, D. (1984). Stochastic relaxation, Gibbs distributions, and the Bayesian restoration of images. *IEEE Transactions on Pattern Analysis and Machine Intelligence, 6*, 721–741.

Gessaroli, M. E., & Folske, J. C. (2002). Generalizing the reliability of tests comprised of testlets. *International Journal of Testing, 2*, 277–295.

Geyer, C. J. (1994). On the convergence of Monte Carlo maximum likelihood calculations. *Journal of the Royal Statistical Society, Series B, 56*, 261–274.

Ghosh, B. K. (1970). *Sequential tests of statistical hypotheses.* Reading, MA: Addison-Wesley.

Gibbons, R. D., & Hedeker, D. R. (1992). Full-information bi-factor analysis. *Psychometrika, 57*, 423–436.

Gilks, W. R. (1992). Derivative-free adaptive rejection sampling for Gibbs sampling. In J. M. Bernardo, J. O. Berger, A. P. Dawid, & A. F. M. Smith (Eds.), *Bayesian statistics 4* (pp. 169–194). Oxford, UK: Clarendon Press.

Gilks, W. R., Richardson, S., & Spiegelhalter, D. J. (1995). *Markov chain Monte Carlo in practice.* New York: Chapman & Hall.

Glas, C. A. W., Wainer, H., & Bradlow, E. T. (2000). Maximum marginal likelihood and expected a posteriori estimation in testlet-based adaptive testing. In W. J. van der Linden, & C. A. W. Glas (Eds.), *Computerized adaptive testing, theory and practice* (pp. 271–288). Boston, MA: Kluwer-Nijhoff.

Green, B. F. (1988). Construct validity of computer-based tests. In H. Wainer & H. Braun (Eds.), *Test validity* (pp. 77–86). Hillsdale, NJ: Erlbaum.

Green, B. F., Bock, R. D., Humphreys, L. G., Linn, R. L., & Reckase, M. D. (1984). Technical guidelines for assessing computerized adaptive tests. *Journal of Educational Measurement, 21*, 347–360.

Guilford, J. P. (1936). *Psychometric methods* (1st ed.). New York: McGraw-Hill.

Gulliksen, H. O. (1950). *Theory of mental tests.* New York: Wiley. (Reprinted, 1987, Hillsdale, NJ: Erlbaum).

Haberman, S. J. (1978). *Analysis of qualitative data. Volume I: Introductory topics*. New York: Academic Press.

Hambleton, R. K. (1986, February). *Effects of item order and anxiety on test performance and stress.* Paper presented at the annual meeting of Division D, the American Educational Research Association, Chicago.

Hambleton, R. K., & Novick, M. R. (1973). Toward an integration of theory and method for criterion-referenced tests. *Journal of Educational Measurement, 10*, 159–170.

Harmon, H. H. (1976). *Modern factor analysis* (3rd ed.). Chicago: University of Chicago Press.

Hastings, W. K. (1970). Monte Carlo sampling methods using Markov chains and their applications. *Biometrika, 54*, 93–108.

Hedeker D., & Gibbons, R. D. (1994). A random-effects ordinal regression model for multilevel analysis. *Biometrics, 50*, 933–944.

Heisenberg, W. (1958). *Physics and philosophy*. New York: Harper & Row.

Holland, P. W. (1990). On the sampling theory foundations of item response theory models. *Psychometrika, 55*, 577–601.

Holland, P. W., & Thayer, D. T. (1988). Differential item performance and the Mantel–Haenszel procedure. In H. Wainer & H. Braun (Eds.), *Test validity* (pp. 129–146). Hillsdale, NJ: Erlbaum.

Holland, P. W., & Wainer, H. (Eds.). (1993). *Differential item functioning*. Hillsdale, NJ: Erlbaum.

Holzinger, K. (1923). Note on the use of Spearman's prophecy formula for reliability. *The Journal of Educational Psychology, 14*, 302–305.

Holzinger, K. J., & Swineford, F. (1937). The bi-factor method. *Psychometrika, 2*, 41–54.

Humphreys, L. G. (1962). The organization of human abilities. *American Psychologist, 17*, 475–483.

Humphreys, L. G. (1970). A skeptical look at the factor pure test. In C. E. Lunneborg (Ed.), *Current problems and techniques in multivariate psychology: Proceedings of a conference honoring Professor Paul Horst* (pp. 23–32). Seattle: University of Washington.

Humphreys, L. G. (1981). The primary mental ability. In M. P. Friedman, J. P. Das, & N. O'Connor (Eds., *Intelligence and learning* (pp. 87–102). New York: Plenum Press.

Humphreys, L. G. (1986). An analysis and evaluation of test and item bias in the prediction context. *Journal of Applied Psychology, 71*, 327–333.

Huynh, H. (1976). On the reliability of decisions in domain-referenced testing. *Journal of Educational Measurement, 13*, 253–264.

Huynh, H. (1990). Computation and statistical inference for decision consistency indexes based on the Rasch model. *Journal of Educational Statistics, 15*(4), 353–368.

Jöreskog, K. G. (1969). A general approach to confirmatory maximum likelihood factor analysis. *Psychometrika, 34*, 183–202.

Jöreskog, K. G. (1974). Analyzing psychological data by structural analysis of covariance matrices. In D. H. Krantz, R. C. Atkinson, R. D. Luce, & P. Suppes (Eds.), *Contemporary developments in mathematical psychology* (Vol. II, pp. 1–56). San Francisco: W. F. Freeman.

Jöreskog, K. G., & Sörbom, D. (1979). *Advances in factor analysis and structural equation models*. Cambridge, MA: Abt Books.

Kelley, T. L. (1924). Note on the reliability of a test: A reply to Dr. Crumm's criticism. *The Journal of Educational Psychology, 15*, 193–204.

Kelley, T. L. (1947). *Fundamentals of statistics.* Cambridge: Harvard University Press.

Kendall, M. G., & Stuart, A. (1967). *The advanced theory of statistics* (Vol. II). London: Charles Griffin.

Kingston, N. (1987). *Feasibility of using IRT-based methods for Divisions D, E and I of the Architect Registration Examination.* Report prepared for the National Council of Architectural Registration Boards. Princeton, NJ: Educational Testing Service.

Kingston, N. M., & Dorans, N. J. (1984). Item location effects and their implications for IRT and adaptive testing. *Applied Psychological Measurement, 8,* 146–154.

Kolen, M. J., Hanson, B. A., & Brennan, R. L. (1992). Conditional standard errors of measurement for scale scores. *Journal of Educational Measurement, 29*, 285–307.

Kuder, G. F., & Richardson, M. W. (1937). The theory of the estimation of test reliability. *Psychometrika, 2*, 151–160.

Lee, W-C., Hanson, B. A., & Brennan, R. L. (2002). Estimating consistency and indices for multiple classifications. *Applied Psychological Measurement, 26*, 412–432.

Levine, M. (1985). The trait in latent trait theory. In D. J. Weiss (Ed.), *Proceedings of the 1982 Item Response Theory and Computerized Adaptive Testing Conference* (pp. 41–65). Minneapolis, MN: Computerized Adaptive Testing Laboratory, Department of Psychology, University of Minnesota.

Lewis, C. (2006). *Validity-based scoring*. Manuscript in preparation. Princeton, NJ: Educational Testing Service.

Lewis, C., & Sheehan, K. (1990). Using Bayesian decision theory to design a computerized mastery test. *Applied Psychological Measurement, 14*, 367–386.

van der Linden, W. J., & Boekkooi-Timminga, E. (1989). A maximin model for test design with practical constraints. *Psychometrika, 54*, 237–248.

Linn, R. E., Baker, E. L., & Dunbar, S. B. (1991). Complex, performance-based assessment: Expectations and validation criteria. *Educational Researcher, 20*(8), 15–21.

Little, R. J. A., & Rubin, D. B. (1987). *Statistical analysis with missing data*. New York: Wiley.

Liu, J. (1993). The collapsed Gibbs sampler with applications to a gene regulation problem. *Journal of the American Statistical Association, 89*, 958–966.

Livingston, S. A., & Lewis, C. (1995). Estimating the consistency and accuracy of classifications based on test scores. *Journal of Educational Measurement, 32*, 179–197.

Loevinger, J. (1947). *A systematic approach to the construction and evaluation of tests of ability* (Psychological Monographs 61, No. 4). Richmond, VA: Psychometric Society.

Lord, F. M. (1967). *An analysis of the Verbal Scholastic Aptitude Test using Birnbaum's three-parameter logistic model* (ETS Research Bulletin RB-67-34). Princeton, NJ: Educational Testing Service.

Lord, F. M. (1971a). The self-scoring flexilevel test. *Journal of Educational Measurement, 8*, 147–151.

Lord, F. M. (1971b). Robbins–Munro procedures for tailored testing. *Educational and Psychological Measurement, 31*, 3–31.

Lord, F. M. (1974). Estimation of latent ability and item parameters when there are omitted responses. *Psychometrika, 39*, 247–264.

Lord, F. M. (1980). *Applications of item response theory to practical testing problems.* Hillsdale, NJ: Erlbaum.

Lord, F. M., & Novick, M. R. (1968). *Statistical theories of mental test scores*. Reading, MA: Addison-Wesley.

Lukhele, R., Thissen, D., & Wainer, H. (1994). On the relative value of multiple-choice, constructed response, and examinee-selected items on two achievement tests. *Journal of Educational Measurement, 31*, 234–250.

MacNicol, K. (1956). *Effects of varying order of item difficulty in an unspeeded verbal test*. Unpublished manuscript, Educational Testing Service, Princeton, NJ.

Mantel, N., & Haenszel, W. (1959). Statistical aspects of the analysis of data from retrospective studies of disease. *Journal of the National Cancer Institute, 22*, 719–748.

Mazor, K. M., Clauser, B. E., & Hambleton, R. K. (1992). The effect of sample size on the functioning of the Mantel–Haenszel statistic. *Educational and Psychological Measurement, 52*, 443–451.

Mazor, K. M., Clauser, B. E., & Hambleton, R. K. (1994). Identification of non-uniform differential item functioning using a variation of the Mantel–Haenszel statistic. *Educational and Psychological Measurement, 54*, 284–291.

McDonald, R. P. (1985). *Factor analysis and related methods.* Hillsdale, NJ: Erlbaum.

McDonald, R. P. (2000). A basis for multidimensional item response theory. *Applied Psychological Measurement, 24*, 99–114.

Messick, S. (1989). Validity. In R. Linn (Ed.), *Educational measurement* (pp. 13–104), Washington, DC: American Council on Education.

Metropolis, N., Rosenblith, A. W., Rosenblith, M. N., Teller, A. H., & Teller, E. (1953). Equation of state calculations by fast computing machines. *Journal of Chemical Physics, 21*, 1087–1092.

Metropolis, N., & Ulan, S. (1949). The Monte Carlo method. *Journal of the American Statistical Association, 44*, 335–341.

Mislevy, R. J. (1986). Bayes modal estimation in item response models. *Psychometrika, 51*, 177–195.

Mislevy, R. J., Beaton, A., Kaplan, B. A., & Sheehan, K. (1992). Estimating population characteristics from sparse matrix samples of item responses. *Journal of Educational Measurement, 29*(2), 133–161.

Mislevy, R. J., & Bock, R. D. (1983). BILOG: Item analysis and test scoring with binary logistic models [Computer program]. Mooresville, IN: Scientific Software.

Mislevy, R. J., & Stocking, M. L. (1987). *A consumer's guide to LOGIST and BILOG* (ETS Research Report 87-43). Princeton, NJ: Educational Testing Service.

Mollenkopf, W. G. (1950). An experimental study of the effects on item analysis data of changing item placement and test-time limit. *Psychometrika, 15*, 291–315.

Monk, J. J., & Stallings, W. M. (1970). Effect of item order on test scores. *Journal of Educational Research, 63*, 463–465.

Moreno, K. E., & Segall, D. O. (1997). Reliability and construct validity of CAT-ASVAB. In W. A. Sands, B. K. Waters, & J. R. McBride (Eds.), *Computerized adaptive testing: From inquiry to operation* (chap. 17, pp. 169–174). Washington, DC: American Psychological Association.

Morris, C. N. (1983). Parametric empirical Bayes inference: Theory and applications with discussion. *Journal of the American Statistical Association, 78*, 47–65.

National Council of Teachers of Mathematics. (1989). *Curriculum and evaluation standards for school mathematics.* Reston, VA: Author.

Neyman, J., & Pearson, E. S. (1928). On the use and interpretation of certain test criteria for purposes of statistical inference. *Biometrika, 20A*, 174–240, 263–294.

Novick, M. R., & Lewis, C. (1967). Coefficient alpha and the reliability of composite measurements. *Psychometrika, 32*, 1–13.

Phillips, A., & Holland, P. W. (1987). Estimators of the variance of the Mantel–Haenszel log-odds-ratio estimate. *Biometrics, 43*, 425–431.

Popper, K. (1987). *The logic of scientific discovery.* London: Hutchinson Publishing Company Ltd. (12th printing).

Ramsay, J. O. (1991). Kernel smoothing approaches to nonparametric item characteristic curve estimation. *Psychometrika, 56*, 611–630.

Ramsay, J. O. (1992). *TESTGRAF: A program for the graphical analysis of multiple choice test and questionnaire data.* Technical report. Montreal, Quebec, Canada: McGill University.

Ramsay, J. O., & Silverman, B. W. (1997). *Functional data analysis.* New York: Springer-Verlag.

Rasch, G. (1960). *Probabilistic models for some intelligence and attainment tests.* Copenhagen: Denmarks Paedagogiske Institut.

Reckase, M. D. (1985). The difficulty of test items that measure more than one ability. *Applied Psychological Measurement, 9*, 401–412.

Reckase, M. D. (1997). The past and future of multidimensional item response theory. *Applied Psychological Measurement, 21*, 25–36.

Ree, M. J., & Earles, J. A. (1991). Predicting training success: Not much more than g. *Personnel Psychology, 44*, 321–332.
Ree, M. J., & Earles, J. A. (1992). Intelligence is the best predictor of job performance. *Current Directions in Psychological Science, 1*, 86–89.
Ree, M. J., & Earles, J. A. (1993). g is to psychology what carbon is to chemistry: A reply to Sternberg and Wagner, McClelland, and Calfee. *Current Directions in Psychological Science, 2*, 11–12.
Resnick, L. B. (1987). *Education and learning to think.* Committee on Mathematics, Science, and Technology Education, Commission on Behavioral and Social Sciences and Education, National Research Council. Washington, DC: National Academies Press.
Ritter, C., & Tanner, M. A. (1992). Facilitating the Gibbs sampler: The Gibbs stopper and the Griddy–Gibbs sampler. *Journal of the American Statistical Association, 87*, 861–868.
Robins, J., Breslow, N. E., & Greenland, S. (1986). Estimators of the Mantel–Haenszel variance consistent in both sparse data and large-strata limiting models. *Biometrics, 42*, 311–324.
Rogers, H. J., & Swaminathan, H. (1993). A comparison of the logistic regression and Mantel–Haenszel procedures for detecting differential item functioning. *Applied Psychological Measurement, 17*(2), 105–116.
Rosa, K., Nelson, L., Swygert, K., & Thissen, D. (2001). Item response theory applied to combinations of multiple-choice and constructed-response items – Scale scores for patterns of summed scores. In D. Thissen & H. Wainer (Eds.), Test scoring (chap. 7, pp. 253–292). Hillsdale, NJ: Erlbaum.
Rosenbaum, P. R. (1988). A note on item bundles. *Psychometrika, 53*, 349–360.
Roznowski, M. (1988). Review of test validity. *Journal of Educational Measurement, 25*, 357–361.
Samejima, F. (1969). *Estimation of latent ability using a response pattern of graded scores* (Psychometrika Monographs Whole No. 17). Richmond, VA: Psychometric Society.
Samejima, F. (1983). Some methods and approaches of estimating the operating characteristics of discrete item responses. In H. Wainer & S. Messick (Eds.), *Principals of modern psychological measurement* (pp. 159–182). Hillsdale, NJ: Erlbaum.
Sands, W. A., Waters, B. K., & McBride, J. R. (Eds.). (1997). *Computerized adaptive testing: From inquiry to operation.* Washington, DC: American Psychological Association.
Sax, G., & Carr, A. (1962). An investigation of response sets on altered parallel forms. *Educational and Psychological Measurement, 22*, 371–376.
Segall, D. O. (2001). General ability measurement: An application of multidimensional item response theory. *Psychometrika, 66*, 79–97.
Shealy, R., & Stout, W. (1993). A model-based standardization approach that separates true bias/DIF from group ability differences and detects test bias/DIF as well as item bias/DIF. *Psychometrika, 58*, 159–194.
Sinharay, S. (2004). Experiences with Markov chain Monte Carlo convergence assessment in two psychometric examples. *Journal of Educational and Behavioral Statistics, 29*, 461–488.
Sireci, S. G., Thissen, D., & Wainer, H. (1991). On the reliability of testlet-based tests. *Journal of Educational Measurement, 28*, 237–247.
Sireci, S. G., Wainer, H., & Braun, H. (1998). Psychometrics: An overview. In P. Armitage & T. Colton (Eds.), *Encyclopedia of biostatistics* (pp. 327–351). London: Wiley.
Sireci, S. G., & Zenisky, A. L. (2006). Innovative item formats in computer-based testing: In pursuit of improved construct representation. In S. M. Downing & T. M. Haladyna (Eds.), *Handbook of test development* (pp. 329–348). Mahwah, NJ: Erlbaum.
Spearman, C. (1904). "General intelligence" objectively determined and measured. *American Journal of Psychology, 15*, 201–293.

Spearman, C. (1910). Correlation calculated with faulty data. *British Journal of Psychology, 3*, 271–295.

Spiegelhalter, D. J., Thomas, A., Best, N. G., & Gilks, W. R. (1994). BUGS: Bayesian inference using Gibbs sampling (Version 0.30). Cambridge, England: Medical Research Council Biostatistics Unit.

Stein, C. (1956). Inadmissability of the usual estimator for the mean of a multivariate normal distribution. *Proceedings Third Berkeley Symposium on Mathematical Statistics and Probability, 1*, 197–206. Berkeley: University of California Press.

Stigler, S. M. (1980). Stigler's Law of Eponymy. *Transactions of the New York Academy of Sciences, 2nd Series, 39*, 147–157.

Subkoviak, M. (1976). Estimating reliability from a single administration of a criterion-referenced test. *Journal of Educational Measurement, 13*, 265–275.

Swaminathan, H., Hambleton, R. K., & Algina, J. (1974). Reliability of criterion-referenced tests: A decision-theoretic formulation. *Journal of Educational Measurement, 11*, 263–268.

Swaminathan, H., & Rogers, H. J. (1990). Detecting differential item functioning using logistic regression procedures. *Journal of Educational Measurement, 27*, 361–370.

Tanner, M. A., & Wong, W. H. (1987). The calculation of posterior distributions by data augmentation (with discussion). *Journal of the American Statistical Association, 82*, 528–550.

Tedeschi, R. G., & Calhoun, L. G. (1996). The Posttraumatic Growth Inventory: Measuring the positive legacy of trauma. *Journal of Traumatic Stress, 9*(3), 455–472.

Thissen, D. (1990). MULTILOG user's guide (Version 6.0). Mooresville, IN: Scientific Software.

Thissen, D. (1991). MULTILOG user's guide (Version 6.1). Mooresville, IN: Scientific Software.

Thissen, D. (1993). Repealing rules that no longer apply to psychological measurement. In N. Frederiksen, R. J. Mislevy & I. Bejar (Eds.), *Test theory for a new generation of tests* (pp. 79–97). Hillsdale, NJ: Erlbaum.

Thissen, D., Chen, W-H., & Bock, R. D., (n.d). MULTILOG 7 – Analysis of multiple-category response data. Retrieved from http://www.assess.com/Software/MULTILOG.htm

Thissen, D., & Steinberg, L. (1984). A response model for multiple choice items. *Psychometrika, 49*, 501–519.

Thissen, D., & Steinberg, L. (1988). Data analysis using item response theory. *Psychological Bulletin, 104*, 385–395.

Thissen, D., Steinberg, L., & Fitzpatrick, A. R. (1989). Multiple-choice models: The distractors are also part of the item. *Journal of Educational Measurement, 26*, 161–176.

Thissen, D., Steinberg, L., & Mooney, J. A. (1989). Trace lines for testlets: A use of multiple-categorical-response models. *Journal of Educational Measurement, 26*, 247–260.

Thissen, D., Steinberg, L., & Wainer, H. (1988). Use of item response theory in the study of group differences in trace lines. In H. Wainer & H. Braun (Eds.), *Test validity* (pp. 147–169). Hillsdale, NJ: Erlbaum.

Thissen, D., Steinberg, L., & Wainer, H. (1993). Detection of differential item functioning using the parameters of item response models. In P. W. Holland & H. Wainer (Eds.), *Differential item functioning* (pp. 67–113). Hillsdale, NJ: Erlbaum.

Thissen, D., & Wainer, H. (Eds.). (2001). *Test scoring*. Hillsdale, NJ: Erlbaum.

Thorndike, R. L. (1951). Reliability. In E. F. Lindquist (Ed.), *Educational measurement* (pp. 569–620). Washington, DC: American Council on Education.

Thurstone, L. L. (1938). *Primary mental abilities* (Psychometric Monograph No. 1). Chicago: University of Chicago Press.

Tierney, L. (1990). *LISP-STAT: An object-oriented environment for statistical computing and dynamic graphics*. New York: Wiley.

Towle, N. J., & Merrill, P. F. (1975). Effects of anxiety type and item difficulty sequencing on mathematics test performance. *Journal of Educational Measurement, 12*, 241–249.

Traub, R. E. & Rowley, G. L. (1991). Understanding reliability. *Educational Measurement: Issues and Practice, 10*, 37–45.

Tucker, L. R. (1958). An interbattery method of factor analysis. *Psychometrika, 23*(2), 111–136.

Wainer, H. (1993). Model-based standardized measurement of an item's differential impact. In P. W. Holland & H. Wainer (Eds.), *Differential item functioning* (pp. 121–135). Hillsdale, NJ: Erlbaum.

Wainer, H. (1995). Precision & differential item functioning on a testlet-based test: The 1991 Law School Admissions Test as an example. *Applied Measurement in Education, 8*(2), 157–187.

Wainer, H. (2000a). Rescuing computerized testing by breaking Zipf's Law. *Journal of Educational and Behavioral Statistics, 25*, 203–224.

Wainer, H. (2000b). *Visual revelations: Graphical tales of fate and deception from Napoleon Bonaparte to Ross Perot (2nd ed.)*. Hillsdale, NJ: Erlbaum.

Wainer, H. (2005). *Graphic discovery: A trout in the milk and other visual adventures.* Princeton, NJ: Princeton University Press.

Wainer, H., Bradlow, E. T., & Du, Z. (2000). Testlet response theory: An analog for the 3-PL useful in testlet-based adaptive testing. In W. J. van der Linden & C. A. W. Glas (Eds.), *Computerized adaptive testing, theory and practice* (pp. 245–270). Boston, MA: Kluwer-Nijhoff.

Wainer, H., Bridgeman, B., Najarian, M., & Trapani, C. (2004). How much does extra time on the SAT help? *Chance, 17*(2), 17–23.

Wainer, H., Brown, L. M., Bradlow, E. T., Wang, X., Skorupski, W. P., Boulet, J., & Mislevy, R. M. (2006). An application of testlet response theory in the scoring of a complex certification exam. In D. M. Williamson, R. J. Mislevy, & I. I. Bejar (Eds.), *Automated scoring of complex tasks in computer based testing* (pp. 169–200). Hillsdale, NJ: Erlbaum, Chapter 6.

Wainer, H., Dorans, D. J., Eignor, D., Flaugher, R., Green, B. F., Mislevy, R. J., Steinberg, L., & Thissen, D. (2000). *Computerized adaptive testing: A primer (2nd ed.)*. Hillsdale, NJ: Erlbaum.

Wainer, H., Dorans, D. J., Flaugher, R., Green, B. F., Mislevy, R. J., Steinberg, L., & Thissen, D. (1990). *Computerized adaptive testing: A primer.* Hillsdale, NJ: Erlbaum.

Wainer, H., & Kiely, G. (1987). Item clusters and computerized adaptive testing: A case for testlets. *Journal of Educational Measurement, 24*, 185–202.

Wainer, H., & Lewis, C. (1990). Toward a psychometrics for testlets. *Journal of Educational Measurement, 27*, 1–14.

Wainer, H., & Lichten, W. (2000). The Aptitude-Achievement Function: An aid for allocating educational resources, with an advanced placement example. *Educational Psychology Review, 12*(2), 201–228.

Wainer, H., Lukele, R., & Thissen, D. (1994). On the relative value of multiple-choice, constructed response, and examinee-selected items on two achievement tests. *Journal of Educational Measurement, 31*, 234–250.

Wainer, H., & Mislevy, R. J. (2000). Item response theory, item calibration and proficiency estimation. In H. Wainer, D. J. Dorans, R. Flaugher, B. F. Green, R. J. Mislevy, L. Steinberg, & D. Thissen, *Computerized adaptive testing: A primer (2nd ed.)* (chap. 4, pp. 65–102). Hillsdale, NJ: Erlbaum.

Wainer, H., Morgan, A., & Gustafsson, J-E. (1980). A review of estimation procedures for the Rasch model with an eye toward longish tests. *Journal of Educational Statistics, 5*, 35–64.

Wainer, H., & Schacht, S. (1978). Gapping. *Psychometrika, 43*, 203–212.

Wainer, H., Sireci, S. G., & Thissen, D. (1991). Differential testlet functioning: Definitions and detection. *Journal of Educational Measurement, 28*, 197–219.

Wainer, H., & Thissen, D. (1996). How is reliability related to the quality of test scores? What is the effect of local dependence on reliability? *Educational Measurement: Issues and Practice, 15*(1), 22–29.

Wainer, H., Wang, X., Skorupski, W., & Bradlow, E. T. (2005). A Bayesian method for evaluating passing scores: The PPoP curve. *Journal of Educational Measurement, 42*(3), 271–281.

Wald, A. (1947). *Sequential analysis.* New York: Wiley.

Wang, M. D., & Stanley, J. C. (1970). Differential weighting: A review of methods and empirical studies. *Review of Educational Research, 40*, 663–705.

Wang, X., Bradlow, E. T., & Wainer, H. (2002). A general Bayesian model for testlets: Theory and applications. *Applied Psychological Measurement, 26*(1), 190–128. Also listed as ETS GRE Technical Report 98-01.

Wang, X., Bradlow, E. T., & Wainer, H. (2004). A user's guide for SCORIGHT (Version 3.0): A computer program built for scoring tests built of testlets including a module for covariate analysis (ETS Research Report RR 04-49). Princeton, NJ: Educational Testing Service.

Weiss, D. J. (1974). *Strategies of adaptive ability measurement* (Research Report 74-5). Minneapolis: University of Minnesota, Psychometric Methods Program.

Weiss, D. J. (1982). Improving measurement quality and efficiency with adaptive testing. *Applied Psychological Measurement, 6*, 473–492.

Weiss, D. J., & Kingsbury, G. G. (1984). Application of computerized adaptive testing to educational problems. *Journal of Educational Measurement, 21*, 361–375.

Whitely, S. E., & Dawis, R. V. (1976). The influence of test context effects on item difficulty. *Educational and Psychological Measurement, 36*, 329–337.

WinBUGS. Retrieved from http://www.mrc-bsu.cam.ac.uk/bugs/welcome.shtml

Wishart, J. (1928). The generalized product moment distribution in samples from a normal multivariate population. *Biometrika, 20A*(32), 424.

van den Wollenberg, A. L. (1982). Two new test statistics for the Rasch model. *Psychometrika, 47*, 123–140.

Wright, B. D., & Douglas, G. A. (1977). Best procedures for sample-free item analysis. *Applied Psychological Measurement, 14*, 281–295.

Wright, B. D., & Stone, M. H. (1979). *Best test design*. Chicago: MESA Press.

Yamamoto, K. (1989). *Hybrid model of IRT and latent class models* (Research Report RR-89-41). Princeton, NJ: Educational Testing Service.

Yen, W. M. (1980). The extent, causes, and importance of context effects on item parameters for two latent trait models. *Journal of Educational Measurement, 17*, 297–311.

Yen, W. M. (1984). Effects of local item dependence on the fit and equating performance of the three-parameter logistic model. *Applied Psychological Measurement, 8*, 125–145.

Yen, W. M. (1992, April). *Scaling performance assessments: Strategies for managing local item dependence*. Invited address presented at the annual meeting of the National Council on Measurement in Education, San Francisco, CA.

Yen, W. M. (1993). Scaling performance assessments: Strategies for managing local item dependence. *Journal of Educational Measurement, 30*, 187–213.

Zenisky, A. L., & Sireci, S. G. (2002). Technological innovations in large-scale assessment. *Applied Measurement in Education, 15*, 337–362.

Zhang, J., & Stout, W. F. (1999). The theoretical DETECT index of dimensionality and its application to approximate simple structure. *Psychometrika, 64*, 213–249.

Zwick, R. (1990). When do item response function and Mantel–Haenszel definitions of differential item functioning coincide? *Journal of Educational Statistics, 15*, 185–198.

Author Index

Subject Index